WHY NOBODY UNDERSTANDS QUANTUM PHYSICS

(And Everyone Needs to Know Something About It)

ABOUT THE AUTHORS

Frank Verstraete is the Leigh Trapnell Professor of Quantum Physics at the University of Cambridge. He has received several prestigious prizes for his research, including the Lieben Prize (also known as the 'Nobel Prize of Austria') and the Francqui Prize. He is among the leading researchers in the world on quantum physics.

Céline Broeckaert is a Romance languages scholar, playwright and author. She co-founded Filmpact, which uses documentary film as a tool for social change and ecological awareness.

Why Nobody Understands Quantum Physics is their first book together.

WHY NOBODY UNDERSTANDS QUANTUM PHYSICS

(And Everyone Needs to Know Something About It)

Frank Verstraete & Céline Broeckaert

Translated by Alice Tetley-Paul & Sue Anderson

MACMILLAN

First published 2023 by Uitgeverij Lannoo

First published in the UK 2025 by Macmillan
an imprint of Pan Macmillan
The Smithson, 6 Briset Street, London EC1M 5NR
EU representative: Macmillan Publishers Ireland Ltd, 1st Floor,
The Liffey Trust Centre, 117–126 Sheriff Street Upper,
Dublin 1 D01 YC43
Associated companies throughout the world

ISBN 978-1-0350-6584-4

Copyright © 2023, Lannoo Publishers. For the original edition.
Original title: *Waarom niemand kwantum begrijpt en iedereen er toch iets over moet weten.*
Translated from the Dutch language.
www.lannoo.com
Translation © Alice Tetley-Paul and Sue Anderson 2025

The right of Céline Broeckaert and Frank Verstraete to be identified as the authors of this work has been asserted in accordance with the Copyright, Designs and Patents Act 1988.

All rights reserved. No part of this publication may be reproduced, stored in a retrieval system, or transmitted, in any form, or by any means (including, without limitation, electronic, mechanical, photocopying, recording or otherwise) without the prior written permission of the publisher.

Pan Macmillan does not have any control over, or any responsibility for, any author or third-party websites (including, without limitation, URLs, emails and QR codes) referred to in or on this book.

1 3 5 7 9 8 6 4 2

A CIP catalogue record for this book is available from the British Library.

Internal illustrations by Céline Broeckaert

Typeset in Minion Pro by Palimpsest Book Production Limited, Falkirk, Stirlingshire
Printed and bound by CPI Group (UK) Ltd, Croydon CR0 4YY

This book was published with the support of Flanders Literature (www.flandersliterature.be).

FLANDERS LITERATURE

MIX
Paper | Supporting
responsible forestry
FSC® C116313

This book is sold subject to the condition that it shall not, by way of trade or otherwise, be lent, hired out, or otherwise circulated without the publisher's prior consent in any form of binding or cover other than that in which it is published and without a similar condition including this condition being imposed on the subsequent purchaser. The publisher does not authorize the use or reproduction of any part of this book in any manner for the purpose of training artificial intelligence technologies or systems. The publisher expressly reserves this book from the Text and Data Mining exception in accordance with Article 4(3) of the European Union Digital Single Market Directive 2019/790.

Visit **www.panmacmillan.com** to read more about all our books and to buy them.

The atoms and their hidden reign,
unseen by the eye,
unknown to the brain.
Our dreams fall short, they cannot hold,
a vision that no mind can mould.
Quantum, a spark in recent years,
met our thought uncertain and unclear.
Yet now we see, as both align:
a deeper glance
rewrites the line.
It breaks the truths we thought we knew,
revealing worlds beyond our view.

CONTENTS

The Physicist's Foreword		1
The Writer's Foreword		5
How to Read This Book		9

PART ONE: MATHEMAGICS

1.	The Unreasonable Effectiveness of Mathematics	13
1.1	How Aristotle was knocked off his pedestal	13
1.2	Mathematics is the language of nature	17
1.3	We know the lion by his claw	20
1.4	Algebracadabra	26
2.	Symmetry	35
2.1	The order of the chaos	35
2.2	When symmetry breaks	40
2.3	Groups: the structure behind symmetries	44
2.4	Drums and atoms	50

PART TWO: QUANTUM

3.	The (Im)probability of a Particle	65
3.1	Max Planck's great tiny quantum leap	65
3.2	Light: wave and particle	76
3.3	The first atomic models	85
3.4	On particles and wave packets	92
3.5	The double-slit experiment	97
3.6	Heisenberg's microscope	100
4.	The First Quantum Revolution	105
4.1	Waves of whimsy	105

4.2	Waves of information	111
4.3	Double slits: the theory	114
4.4	Quantum tunnelling	116
4.5	Matrix mechanics	118
4.6	Beauty is truth, truth is beauty	121
4.7	The spin becomes a qubit	129
5.	Quantum Philosophy	139
5.1	Quantum spiritualism	139
5.2	Entanglements	141
5.3	Niels versus Albert	142
5.4	The EPR paradox	146
5.5	Schrödinger sends his cat	152
5.6	The cat rings the Bell	156
5.7	Contextuality	162
6.	One, Two, Many	167
6.1	The indistinguishability of particles	168
6.2	Hotel Hilbert	172
6.3	Atoms and molecules	183
6.4	Hard matters	200
6.5	Quantum colour	206
6.6	On Bose, Einstein and lasers	210
7.	Pudding and Quark	221
7.1	Subatomic physics: the experiments	221
7.2	Subatomic physics: the theory	243
7.3	We are all made of stars	262
8.	More Is Different	269
8.1	Emergence	269
8.2	Renormalizing	278
8.3	Superconductivity	281
8.4	The discovery of perfection	287
9.	The Second Quantum Revolution	295
9.1	Quantum metrology	296

9.2	Quantum simulation	298
9.3	Quantum information	301
9.4	Quantum complexity	305
9.5	Quantum computers	313
9.6	Quantum in error	316
9.7	Quantum 2.0: entangled particles	321
	Epilogue	335
	Acknowledgements	339
	Glossary	341
	Further Reading	351
	Index	354

THE PHYSICIST'S FOREWORD

This book is the result of a clash of cultures: between the scientific quest to 'mathematize' the world, gradually unravelling its mysteries, and the beauty and amazement we encounter in the realm of transcendent art; between uncovering symmetrical patterns in our material world and the simplicity and splendour of Schubert's music; between the inherent unpredictability of a quantum system and the deeply conflicting emotions found in Hermann Hesse's *Narcissus and Goldmund*.

The point is: there is no clash. Schrödinger's equation belongs just as much to our cultural canon as Beethoven's Ninth Symphony. Understanding how quantum mechanics gives colour to our world is just as delightful as admiring the splendid colours in Klimt's paintings. Understanding how symmetry serves as a central organizing principle, preventing us from shrinking into a mere pea, is as humbling as standing before the majesty of the Grand Canyon. In essence, physics aspires to achieve what Michelangelo envisioned when he gazed upon the marble block that would become David: to observe and describe nature in such a way that all excess material can be chipped away, leaving only the true essence.

However, there is a fundamental difference: while contemporary art and art from hundreds or thousands of years ago is equally valuable and arouses equally strong emotions, they do not truly build upon each other. Art constantly reinvents itself and has to be (or not be) original. This does not apply to physics, where one theory builds on another in a seamless progression. Newtonian mechanics gave way to Einstein's theory of relativity, which in turn evolved into quantum field theory; the quantum realm of the subatomic unveiled profound mysteries about chemistry, matter and the workings of the stars; alchemy transformed into an experimental science capable of turning mercury into gold. And this is precisely what makes the story of quantum physics so captivating and so fascinating to tell: it is a story of countless individuals, each building on each other's ideas, igniting revolutions that fundamentally reshape our understanding of the world. It's the story of how a handful of foundational quantum concepts enable us to comprehend an endless array of natural phenomena.

The premise of this book is simple: quantum physics is far from incomprehensible. On the contrary, it can be made accessible to all. By understanding a few foundational concepts – such as symmetry, the exclusion principle and the uncertainty principle – anyone can connect with the atomic world, which forms the basis of numerous technological applications we use every day. The inherent complexity and counterintuitive nature of physics must not be used to mystify it. That's precisely the opposite of what a popularizing book should seek to do. Nor should readers be expected to follow every line of reasoning – that is only possible in the precise language of mathematics. Instead, the philosophy of this book is to encourage readers to focus on the ideas themselves, as these are much more intuitive and important than the technical deductions behind them. We do not intend – nor is it necessary – to explain the full mathematics of quantum physics in a book without formulas. What we aim to do is something else entirely: to inspire readers to see and feel new things, and to view the world from a fresh, quantum-inspired perspective.

Quantum physics works, and we are on the verge of a second quantum revolution – one that will profoundly transform our technology. Anyone who is curious about the fundamental workings and beauty of our world should have at least a basic understanding of quantum physics. That is the mission of this book.

THE WRITER'S FOREWORD

At school, no one ever managed to explain the purpose of mathematics to me. 'Why do we have to learn this, Miss? It's not like we're ever going to need that later, right?' The same old refrain, you know how it goes. Cross and frustrated, I would storm off, finding solace in the poster pinned to the wall above my bed: 'Do not worry about your difficulties in mathematics. I can assure you that mine are still greater.' Signed: Albert Einstein. Those of us not blessed with the divine gift of numbers resorted to borrowing answers from the clever classmate sitting next to us. By the time the holidays rolled around, all was forgiven – and promptly forgotten. And so, I earned my marks through book reviews, essays and presentations, where at least my knack for language served me well.

What actually frustrated me wasn't just my struggle with mathematics, but my inability to understand *why* I struggled with it. Today, I've come to realize that maths isn't so different from prose; it's simply another form of expression, another language. It has its own rules, its own rhythm, its own poetry. Mathematics is the ultimate workout for our capacity to think abstractly, to ask the right questions, to (learn to) solve problems and to discover connections. Consider Newton: with one and the same formula, he explained both why an apple falls to the ground and why planets orbit the sun – a perfect testament to the elegance of mathematical thought.

Is it really that necessary, then, to understand how everything fits together? No, not really. Is it invaluable? Absolutely. Our point is this: quantum physics is an undeniable part of our culture, to the same extent as literature, music, theatre, film and all that jazz. Because culture equals knowledge. It's the translation of how we evolve as humans, how we engage with our history, how we relate to the infinitely large and the infinitely small, and to everything we do and everything we have no control over. We derive our identity from culture. And when we think of identity, we inevitably think of history. Behind every turning point in history stands not just a strong man or woman, but also a strong idea. Nearly all of those ideas came about as a result of new insights in the natural sciences – think of the Enlightenment, industrialization, automation, globalization, digitization, et cetera. Therefore, it is our hope that thanks to (or: in spite of) this book, students will be inspired to study the sciences. Especially girls. Not because of their gender – though naturally also because of it. For girls and women are remarkably gifted in mathematics and science. This book provides hard proof of that.

This reminds me of my master's thesis, which focused on the translation of René Char's *Feuillets d'Hypnos* into Italian. I had come across that little book one day, rather by accident; but, like the translator, I didn't initially 'get it'. The French felt unnecessarily difficult, almost impenetrable. And yet. Char's aphorisms haunted me. I couldn't quite shake the feeling that, in every text, he was reaching for something essential, something 'beyond'. For months, I wrestled with the words, searching for meaning, until suddenly, one *aha* moment followed another. So *that's* what he meant! Through the most mundane occurrences in daily life, I began to see connections I hadn't noticed before, unravelling what he was truly trying to convey. Slowly, I found myself internalizing those aphorisms. I began to see the world through their lens, peeling back layers to reveal their core meaning – and, in the process, finding my own. And that's exactly what I hope you will experience through these pages.

I wouldn't go so far as to claim that, since writing this book, I now roam the supermarket aisles in search of a pot of (sugar-free) quark with mathematical formulas running through my mind. But as I stand at the self-checkout scanning that pot of quark, I find myself thinking of the chapter on lasers. When I text home to say I'll be a little late, I recall the section on transistors. And when I reflect on the fact that my mother has been cancer-free for over twenty years, I'm overjoyed that MRI scanners exist and that I can see her smiling on my computer screen as we talk via Zoom, each in a different continent. Everything takes on a deeper meaning when you understand how it all works. And whether we owe it to a poet or a quantum physicist, what truly matters is the value and significance we attribute to the world around us – the way it reshapes our thinking and, ultimately, helps us grow.

One final thought on my collaboration with Frank. I thought our ways of thinking couldn't be more different: mine felt much more instinctive, less linear, not bound by the black-and-white logic that is (seemingly) inherent in science and mathematics. I tend to live, work and think in a very intuitive way. For me, life is a succession of surprises and encounters, with plenty of scope for serendipity. But as I listened to Frank, I realized that we're not so different after all. Actually, we think the same way and share the same craving for beauty. We are equally chaotic, equally creative and equally driven by a sense of urgency and the desire to share what we have and what we love with others. We just express this in completely different ways. And that's exactly what made working together so fun – and so essential. It compelled us to listen even better to each other, to step into each other's world, to try even harder to understand. It's a matter of sensing and complementing each other. Of asking the right questions. Drawing the right connections. Is the professor foolish because he can't explain it? Am I foolish because I don't understand it? Perhaps! But the difference lies here: Frank thinks and speaks in mathematics; I think and speak in words.

At the same time, it's all relative. There's mathematics, quantum physics, formulas and entanglements, but there are other things in life that are far more complex. In daily life, we must learn to resolve conflicts, to find words for our feelings and thoughts. Here, too, we search for language, for understanding, for clarity. What we wrote here not only had to be correct and clear, but also beautiful. Knowledge should be something you *want* to pass on – something infectious. I hope that message resonates with you and that you don't walk away feeling cross or frustrated. And if the subject still feels a little out of reach after reading this book, remember: that's perfectly normal. At the very least, I hope it's been fun to read. And if some things haven't clicked yet – don't worry, that *aha!* moment will come.

Céline Broeckaert and Frank Verstraete
Ghent, Cambridge and Nyons, June 2023

HOW TO READ THIS BOOK

This is not a physics book. It's a book *about* quantum physics. After all, the essence of quantum physics is not mathematics – it's the ideas behind it that matter most. That's why we have de-ciphered our story as much as possible by stripping it of formulas and mathematics. Quantum logic is notoriously strange – baffling, even – and it can feel downright impossible to grasp at times. But that's okay. You don't need to master every detail to appreciate the elegance and beauty of nature's laws. Bear in mind that even quantum physicists don't fully grasp quantum physics. They've simply learned to live with it; they've developed a certain intuition for it. So, let the waves carry you away. Think of it like listening to music: you don't have to understand the mathematical underpinning to enjoy it. Each chapter develops a variation on the quantum theme. You can savour it, even if you can't untangle every nuance.

The structure of the book

This book is about quantum physics: how it emerged in the early twentieth century, evolved into the greatest revolution in our understanding of matter and how it underpins much of the technology we depend on. Yet, like all branches of science, quantum physics is just one chapter in a never-ending story. Where does such a story begin? How far back must we go to uncover its roots? We've chosen to begin in the sixteenth century with Simon Stevin, one of the first geniuses to challenge scientific dogmas and let experiments guide him towards surprising, counterintuitive truths. From there, the story of quantum physics unfolds like a relay race, with knowledge passed from one generation to the next. Step by step, it grew into an essential part of how we perceive the world and played a key role in shaping our everyday lives. That's the thread connecting everything in this book.

Alongside the main text, this book also contains two other types of text:

FOR THE RECORD: these are the text sections marked with a vertical line. They bring a lively twist to the text.

FOR THE AFICIONADOS: these are the framed sections, for those who want to dig a bit deeper and embrace the figures and formulas. Rest assured, these sections are not necessary in order to follow the main text.

PART ONE

MATHEMAGICS

IN A NUTSHELL

> Physics stands or falls on the results of experiments.

> Mathematics: the language of nature (and unreasonably effective).

> Physics is about ideas (not about mathematics).

> *Dramatis personae*: Simon Stevin, Galileo Galilei, Isaac Newton, William Hamilton.

ONE

THE UNREASONABLE EFFECTIVENESS OF MATHEMATICS

1.1 How Aristotle was knocked off his pedestal

Simon Stevin

It all began in the sixteenth century, in the year 1586, with what seemed like a pointless experiment. The scientific playground: the Dutch city of Delft. Two visionary minds, the Flemish scientist Simon Stevin (a Bruges-born mathematician, physicist, proto-engineer and master of doubt) and his friend Jan Cornets de Groot (whom Stevin called 'the most diligent seeker of Nature's hidden secrets'), climbed to the top of the Nieuwe Kerk tower armed with two hefty lead balls, one ten times heavier than the other. From a height of thirty feet (about nine metres), they dropped the balls in perfect synchrony.

Below, a third man – firmly planted on the ground, entrusted with the seemingly simple yet crucial task of observing – confirmed with his own eyes (and ears) that the two lead balls landed on the wooden platform at exactly the same moment. There was unmistakably only one, singular *thud* to be heard. By the time he stormed down the steps, breathless and drenched in sweat, Stevin's conclusion was ready: Aristotle's celebrated theory of gravity was false! For nearly 2,000 years (that is, ever since Aristotle), everyone had taken it for granted that heavier objects fall faster than lighter ones.[1] But the verdict was unambiguous. They had all been wrong!

The study of falling bodies marked the beginning of an unprecedented scientific revolution. And so, our story begins. In the centuries that followed, the groundwork was laid: reason illuminated minds, knowledge expanded and the world moved steadily toward the birth of quantum physics.

Stevin, often called the 'Da Vinci of the Low Countries', used the opportunity to delicately point out the weight of his experiment: no matter how elegant and logical – or even romantic, if you like – a theory might seem, and no matter how much it appeals to intuition, if an experiment proves it wrong, it's worthless. The world of abstract ideas was politely asked to step aside in favour of empirical research

---

1 Provided we abstract certain details that distract from the essence. The objects being tipped over the parapet must be heavier than air; but they shouldn't fall too fast, either, or air resistance becomes too significant. It wouldn't work with a helium balloon, as that falls upwards. In a vacuum, on the other hand, everything falls at the same speed regardless.

and the harsh clarity of reality. Today, that might seem fairly obvious, but at the time, it marked a monumental shift from tradition. It also explains why such contrary viewpoints were initially met with fierce opposition. Of course, you can only conduct tests if you have the necessary instruments. The rapid acceleration of scientific progress in the seventeenth century was a direct result of inventions such as the telescope and the microscope. The more precise the measurements became, the more experiments revealed new insights and dismantled the once-unshakeable theories of the past.

This idea will weave its way throughout this story. The history of science is a perpetual game of ping-pong between theory and experiment, between thinking and testing. Ultimately, it is always experiments – not reason or gut instinct – that decide whether a new theory is needed. Scientists aren't preoccupied with who discovered something first. The question that keeps them up at night is: what scientific law explains what I observe? And can that law predict the outcome of future experiments? This is the cornerstone of the scientific method – the only true way to do science. In the end, our intuition is merely shaped by daily experiences with the relatively large – the visible world, so to speak. But that intuition falters the moment we delve into the microscopic realm. An atom, which roughly consists of a nucleus surrounded by electrons, is not at all akin to a miniature sun with planets orbiting around it. Still, a solid understanding of the macroscopic world can absolutely shed light on the microscopic, and vice versa.

If we want to delve into quantum physics, we must adopt this radical mindset of Stevin. It was this same unrestrained spirit that drove the pioneers of quantum physics,[1] 350 years after Stevin, to

---

1 Quantum physics is a synonym of quantum mechanics. We use both terms interchangeably.

turn the entire field of physics on its head. Since there were certain experiments and mysteries that couldn't be deciphered, not with the best will in the world, and certainly not with classical physics, Werner Heisenberg and Erwin Schrödinger approached the problem with open minds and a willingness to rethink everything. The result? A groundbreaking new theory that delivered the answers classical physics failed to provide. This burst of innovation in science sparked a cascade of new questions, astonishment, daring ideas and ultimately the foundational laws of quantum physics. It challenged us to accept some truly strange truths: particles can behave like waves, and they can exist everywhere and nowhere at once. It all depends on how you look at them, and what and how you measure. And yes, quantum physics is packed with paradoxes, but that's part of its charm. In exchange, it gives us an understanding of the tiniest building blocks of matter and a completely fresh perspective on reality, and it provided the essential framework for the incredible technological advances of the twentieth century.

Why did we start this book with Simon Stevin? Not only because he's the great-grandfather of the scientific method, but also because he championed the use of everyday language (rather than Latin) as the ideal medium for doing science. In that spirit, we too will lean on accessible language to achieve our goal: demystifying quantum physics and explaining its fundamental principles to a broad audience. Lastly, we get the ball rolling with Stevin because the protagonist of the next part of the story, Galileo Galilei, owes an undeniable debt to this visionary from Bruges. Credit where credit is due.

GROPING AROUND IN THE DARK[1]

Once upon a time, a drunken man stumbled back and forth beneath a streetlamp, scanning the ground for his keys. A passer-by, noticing his plight, stopped to help with the search. After a while, when their efforts turned up nothing, the passer-by asked if the man was sure he had lost his keys in that spot. The drunk shrugged and slurred, 'No. But at least there's some light here.' The story carries a deeper truth: nature may overwhelm us with mysteries, but it's our job to discover the laws – the keys – that unlock them. It's only natural to look where the light shines, because that's where we can see. But as we'll come to understand, sometimes we must grope around in the dark to truly find what we're looking for.

1.2 Mathematics is the language of nature

Once there was a young man who loved to study but wasn't particularly fond of attending church. One day, while Galileo Galilei (1564–1642) was supposed to be dutifully murmuring psalms, his attention wandered to a lamp, suspended on a chain, swinging gently back and forth from the ceiling of Pisa Cathedral.[2] Lacking a proper timepiece, he improvised, using the rhythm of his own heartbeat to measure how long it took for the lamp to complete each swing. With this enlightening little experiment, Galilei discovered the

---

[1] See Giorgio Parisi, *La chiave, la luce e l'ubriaco* [*The key, the light and the drunkard*], Di Renzo Editore, 2006.

[2] Modern biographers query the authenticity of this anecdote, as it is possible that the lamp in question was not hung in the cathedral until two years after Galilei's visit (in 1585).

counterintuitive law behind a swinging object: no matter how wide the swing of a pendulum, the time it takes to complete each arc remains exactly the same.

Galileo Galilei would not have been Galileo Galilei if he hadn't taken ample inspiration from his contemporaries. But with his own 'drop test 2.0', the father of modern science took things a step further than Stevin. He set his sights on the elusive concept of time and its intervals. The challenge? He couldn't quite put it into words. To untangle the mechanics of something as abstract as 'time', Galilei had to turn to the one tool that could make sense of it: mathematics.[1]

One crisp, starry night, nestled among the cypress trees of idyllic Pisa, Galilei found his conviction growing stronger. As he peered at the sky through his telescope, an avant-garde toy of its time, a sudden thought crossed his mind: what if I spotted another civilization out there? What language could I use to speak to them? Stevin's Dutch and Galilei's own Italian wouldn't get him any closer. *E allora . . . ? Matematica!* Though he hadn't yet cracked the formula for his swinging objects, the realization that mathematics was the universal key – to both unlocking the mysteries of nature and bridging the gap with extraterrestrials – was itself a breakthrough. As Galilei so memorably put it: 'The book [i.e., nature] [is] written in the language of mathematics [. . .] and without that [knowledge] it is impossible to understand even a single word [of the book].'

Here's the essence of it: the laws of nature exist independently of us. Humanity didn't invent mathematics to explain nature – mathematics *is* the language of nature. And every so often, experiments compel scientists to expand that language's vocabulary. This is why the outcome

---

[1] The Dutch term for mathematics, *wiskunde*, was coined by Simon Stevin himself. It stems from *wis-const*, or 'the art of the wise', capturing the essence of anything rooted in numbers and certainty. Stevin also introduced the Dutch words for physics and geometry: '*natuurkunde*' ('the art of nature') and '*meetkunde*' ('the art of measuring').

of an experiment is objective and unchanging, always leading to the same result, no matter where or when. Whether on Earth or Mars. A mathematical model of the world doesn't just help us describe what we see – it allows us to make predictions and in turn to falsify[1] those predictions on the basis of experiments. That process of testing, refining and understanding – that's physics, in its purest form.

Galilei 'mathematized' physics, pulling it away from the realms of philosophy and religion, a bold move that earned him the ire of the church and plenty of sceptics. Centuries later, quantum physics would take this separation to its extreme. Thanks to (very) abstract mathematics, we can now predict and explain nearly everything in nature. Everything in the world is expressed through mathematical functions. The sound slipping into your ears: functions! The light landing on your retina: functions! The path a particle travels over time: functions! The warmth spreading through a too-crowded cafe: functions! The probability that a measurement will give you the answer you're hoping for: functions! Even the butterfly in the Amazon flapping its wings and somehow nudging the weather in your town: functions! Functions are the language of the universe, describing how everything depends on something else.[2] Yet, for all its elegance, mathematics can only be fully understood, and made tangible, through words. Because ultimately, physics isn't about maths; it's about ideas. And behind every transformative idea is a person – a giant, if you will. The next giant in our story is Isaac Newton.

---

1 In physics, you can't prove that a theory is right. You can only show that it is wrong. This is what we mean by 'falsifying'. Just because an experiment works doesn't guarantee your theory is correct. It could just be down to chance.
2 For a forceful argument on functions, check out Thomas Garrity's lecture series, 'On Mathematical Maturity'.

1.3 We know the lion by his claw

The year was 1665. A pandemic had brought Europe to a standstill, universities shut their doors and Isaac Newton (1643–1727), then a Cambridge student in his early twenties, retreated into quarantine at his family's manor in Woolsthorpe, about seventy miles (roughly 110 km) north of the city. What followed was nothing short of an *annus mirabilis* for Newton. Never before had anyone introduced so many new concepts and gained so many insights into physics in such a short time as Newton did – a man who, by the way, never travelled further than London, not even to see the sea (despite his enormous fascination with waves); a man who was said never to have been loved by any woman, not even by his mother, and perhaps even because of his mother, who left him in the care of his grandmother at the age of three. Whether this lack of warmth contributed to Newton's notoriously difficult personality is up for debate, though such details are not really relevant here.

At the time, the biggest conundrum in physics was how to describe the orbits of the planets. Or rather, the glaring lack of a way to do so. The mathematics of the day simply wasn't up to the task. Newton, ever undaunted, decided to fix this by inventing a brand-new branch of mathematics: calculus. With differential and integral calculus, Newton provided a tool that could do something revolutionary: if you know the position and speed of particles, and the forces acting between them, you can use calculus to determine both their past and their future. In essence, everything in the universe boils down to movement and change. Everything influences everything else and everything evolves over time: clocks tick, grass grows, planets revolve, electrons spin, cats grin.

Using calculus, Newton cracked what had been the biggest

problem in physics. With a single, unified formula, he described the orbits of planets and explained why apples fall from trees. The same laws applied to both phenomena. This was pretty spectacular as, prior to Newton, the science of celestial motion and earthly physics were thought to belong to completely separate domains. Thanks to calculus, even Stevin and Galilei's pendulums and falling objects could (finally!) be expressed in precise mathematical terms. In doing so, Newton accomplished what Galilei couldn't: he wielded mathematics to describe the behaviour of accelerating bodies, unifying the heavens and the Earth under the same set of physical laws.

Isaac Newton

NEWTON VERSUS LEIBNIZ

Nine years after Newton's *Philosophiae Naturalis Principia* (better known as *The Principia*) was published, a competition was announced to challenge the world's sharpest scientific minds. The men behind the plan to find out who was truly the great master of calculus were the mathematicians Johann Bernoulli and Gottfried Wilhelm von Leibniz. They posed a fiendishly difficult problem, to be solved within six months: what is the optimal shape of a marble run to ensure that a marble rolls from point A to a lower point B in the shortest time possible? When Newton received the challenge, it must have unleashed a torrent of brilliance within him, because he solved it in just twelve hours. However, wary of his rival Leibniz, Newton submitted his solution anonymously. It was a futile effort. Newton's distinctive mathematical prowess gave him away immediately. Bernoulli could not have put it any more aptly: 'We know the lion by his claw.'

While the rivalry between Newton and Leibniz may appear, at first glance, to be a petty squabble over 'who was first', the stakes were anything but trivial. The debate concerned nothing less than the invention of calculus, one of the most transformative discoveries in scientific history. Today, historians largely agree on a nuanced conclusion: Newton and Leibniz developed calculus independently of each other. But Newton got there first.

Newton's laws of motion and calculus introduced the concept of a deterministic universe, aligning with the philosophical framework of his near-contemporary, René Descartes. This model left no room for chance and firmly dismissed randomness. For Newton, the universe was as orderly and predictable as the ticking of the clock in his kitchen. Two o'clock would always be followed by half-past two, sixty minutes would always make an hour and tomorrow's twenty-four hours would unfailingly mirror today's. In Newton's eyes, the universe operated in much the same way: if you know the position and velocity of every particle – be it stars, planets, moons or apples – at a given moment, you could, with absolute precision, calculate where they would be at any future point in time and trace their trajectories back into the distant past.

From all of the above, we can confidently conclude that Newton is the father of all classical physics. However, it will eventually become clear that Newton's classical physics has its limitations. It struggles to describe the very smallest scales, and that's where the laws of quantum physics will take over. Similarly, it falls short when dealing with the very largest scales or objects approaching the speed of light, gaps that Einstein would later fill with his theory of relativity. Even Newton's deterministic worldview would face significant challenges. This is because, in quantum physics, chance isn't just a consequence of our ignorance; it plays a central role. At the quantum level, everything exists by virtue of randomness and probability. Remarkably, quantum physics used Newton's own invention, calculus, to undermine the theories he built with it. No new mathematics was required for this shift. Instead, the same calculus formulas were simply reinterpreted in a fundamentally different way, this time through the lens of wave functions and probabilities – two concepts, as the chapters ahead will reveal, that form the core thread of this story.

What quantum physics also shares with calculus is its remarkable *unreasonable effectiveness*. Initially developed to describe how electrons orbit a nucleus, quantum physics turned out to have far broader applications than anyone could have anticipated. Eugene Wigner (1902–1995) eloquently captured this sense of wonder at the 'unreasonable effectiveness' of mathematics in the natural sciences: 'It is difficult to avoid the impression that a miracle confronts us here, quite comparable in its striking nature to the miracle that the human mind can string a thousand arguments together without getting itself into contradictions, or to the two miracles of the existence of laws of nature and of the human mind's capacity to divine them.'

Simon Stevin had championed Dutch as the language of science. Galilei replaced everyday language with mathematics and, with Newton's calculus, mathematics reached an absolute pinnacle. Newton was, in short, the first true scientist – or perhaps more aptly, the last of the sorcerers. Before his time, people mostly operated on intuition, mixing and matching ideas and methods in a haphazard way. With a fair share of patience and a bit of guesswork, they occasionally stumbled upon explanations for everything (okay, not quite *everything*). Then suddenly, the planets were orbiting the sun – not the other way around. Except no one could explain *why* that was so. The 'why' stopped dead at a question mark. Although Galilei and Stevin had introduced the scientific method, it was Newton who gave it structure and substance. His genius lay in defining a methodical approach to understanding nature, which required new insights, fresh tools and a mindset that would take generations to fully develop. Our next giant, the brilliant astronomer and mathematician Sir William Rowan Hamilton, was ready to take up that challenge.

THE APPLE

'[Newton] is our Christopher Columbus. He has led us to a new world, and I would like to travel there.' When we think of Newton, it's almost impossible not to picture the apple. Yet the famous apple story wasn't born under a tree but in the creative mind of writer, philosopher and fervent Newton admirer François-Marie Arouet, better known as Voltaire (1694–1778), the author of the quote above. While Voltaire's tale of the falling apple owes more to imagination than to history, it proved an ingenious way to humanize Newton and make his revolutionary ideas more accessible to the broader public.

À propos, if you've ever wondered why some of Voltaire's works occasionally display striking scientific depth, the answer lies in the embrace of Émilie du Châtelet (1706–1749). For many years, Voltaire and du Châtelet lived in the best of worlds, spending their days immersed in love, literature, theatre and rigorous study. But why does Émilie du Châtelet deserve special recognition? Because she was the first female scholar of the modern era, a physicist and mathematician who tackled some of the most complex problems with astonishing clarity and ingenuity. She refused to let the male-dominated world of science exclude her from Newton's ideas. But most notably, she translated Newton's monumental *Principia* from Latin into French. And not just translated, excuse us: she refined, expanded and supplemented it, ensuring the work was as precise as it was comprehensive, particularly regarding the concept of energy. Du Châtelet demonstrated that energy cannot be created or destroyed, only transformed. She also settled a lingering dispute between Newton and Leibniz by proving that kinetic energy is proportional not to velocity (Newton's view) but to the square of velocity (Leibniz's). While Newton may have triumphed in the calculus wars, this time, Leibniz had it right. Du Châtelet's insights on energy would later become central to the work of William Hamilton and Joseph-Louis Lagrange, whose work would provide the mathematical foundation of quantum physics.

Du Châtelet was over forty when she became pregnant (by a man who was not Voltaire). Given her age, she grew increasingly anxious about the risks and raced to complete her translation. The child was born healthy, but du Châtelet died shortly after, leaving Voltaire devastated. Before leaving France, he mustered all his courage (and honour) to have her *Principes mathématiques de la philosophie naturelle d'Isaac Newton* published posthumously. To this day, it remains the definitive French edition of Newton's work. For the Epicureans among us, we recommend du Châtelet's *Discours sur le bonheur* (1779). Oh, lest we forget, a crater on Venus was named after her. Though we prefer to mark our appreciation in this book.

1.4 Algebracadabra

One fine day, Sir William Rowan Hamilton (1805–1865) fell in love. But not just in love, oh no. He fell in love as only an astronomer could: *to the moon and back*. On an otherwise unremarkable 16 October in 1843, he stepped out of his study to take a stroll in the precious company of his beloved. As they wandered along, navigating unanswered questions such as what might be for dinner that evening, Hamilton arrived at a bridge spanning Dublin's Royal Canal. There, in a sudden moment of blinding brilliance, he was struck by an insight so profound that he felt compelled not to carve his beloved's name into the stone but rather his highly poetic equation[1] $i^2 = j^2 = k^2 = i \cdot j \cdot k = -1$.

'Quaternions! Take that, my love!' he declared triumphantly. 'This is how you extend complex numbers into four dimensions! Away with commutativity!' Lady Hamilton, however, simply nodded, feeling slightly passed over. Setting aside this minor romantic misstep, Hamilton's spontaneous act of mathematical graffiti would go on to lay the cornerstone of linear algebra and the study of matrices. The fact that, even today, throngs of enthusiasts gather

---

1 To clarify, a dot signifies multiplication.

each 16 October to retrace his steps during the Hamilton Walk is a testament to how much the scientific world owes to this one moment of genius.

Hamilton's bridge over the Royal Canal

Before we go any further, we need to take a brief but necessary detour into the world of complex numbers and quaternions. One of the reasons mathematics and quantum physics can feel so intricate is their reliance on these unusual numbers. Most of us 'normal' people stick to real numbers, like 9 or 14 or −65 or 3.14 or −√2. Square any of those, and you'll always end up with a positive result. After all, a number multiplied by itself can't help but be positive. People with a slightly more mathematical mindset also work with imaginary numbers, such as the number *i*, which is the square root of −1 (so *i* . *i* = −1 and 10 . *i* is the square root of −100).

Complex numbers combine these two worlds: they have one real and one imaginary component. They might seem strange at first, but they prove incredibly useful for tackling complex problems. Before the advent of quantum physics, complex numbers were treated as a mathematical oddity, a clever tool to solve differential equations more efficiently and elegantly. Quantum physics, however, changed all that. Complex numbers shed their reputation as mere mathematical tricks and became indispensable. At the heart of it all is a remarkable fact: nature itself cannot be fully described without complex numbers.

Back to our main thread: Hamilton. He didn't just stop at complex numbers, he took a leap forward with quaternions, laying the mathematical foundation for quantum physics. And that mathematical basis is, yes, pretty complex. Unlike 'ordinary' complex numbers, which consist of two components, quaternions have four: 1 (a real number), i (an imaginary number), and two additional components, j and k. Their most notable property? Quaternions are non-commutative, meaning the order of multiplication matters: $i \cdot j$ is not the same as $j \cdot i$. To put it in everyday terms: filling a swimming pool before jumping in is not the same as jumping in first and filling the pool afterwards. The correct relationship is $i \cdot j = -j \cdot i$.

However, quaternions *are* associative, which means that the grouping of terms doesn't affect the result: $a \cdot (b \cdot c) = (a \cdot b) \cdot c$. In simpler terms, boiling and peeling an egg before eating it is no different from boiling the egg first, peeling it second and then eating it. This dual nature of quaternions (non-commutative but associative) opened up entirely new mathematical possibilities, ultimately forming the backbone of quantum physics.[1]

It soon became clear that this way of working with 1s, *i*s, *j*s and *k*s was a bit complicated after all. It needed to be simpler. And it *could* be made simpler; by invoking matrices. A matrix is like a big chequerboard filled with complex numbers, arranged in rows and columns. It might look like the image opposite.

---

[1] For the quantum aficionados: eighty years later, quaternions will crop up again in the context of the famous Pauli spin matrices, which describe the spin of an electron.

$$A = \begin{bmatrix} a_{11} & a_{12} & \cdots & a_{1n} \\ a_{21} & a_{22} & \cdots & a_{2n} \\ \vdots & \vdots & \ddots & \vdots \\ a_{m1} & a_{m2} & \cdots & a_{mn} \end{bmatrix}$$

A matrix with m rows and n columns

Switching to matrices opened up a world of possibilities, allowing non-commutative structures to be applied to far more interesting systems. Where Newton's theory enables scientists to describe how a single particle evolves under the influence of applied forces, matrices allow them to do the same, but for entire systems of particles at once. These so-called 'many-particle systems' consist of countless particles, all exerting forces on each other. The numbers in the matrix show how strongly the particles attract or repel one another and predict how the entire system will evolve over time. A key concept here is *eigenfrequencies*.[1] Every matrix has its own set of eigenfrequencies – its natural resonances – that reveal the vibrations of a many-particle system. And since every system vibrates, it's these frequencies that dictate how particles move.

COMMON-OR-GARDEN QUATERNIONS

What's the point of all these abstract concepts? As it turns out, quaternions and matrices are the unsung heroes behind the lightning-fast image processing that powers smartphones,

---

[1] The prefix 'eigen-' is a German word, meaning 'characteristic'. Eigenfrequencies encode the characteristic properties of the system under study.

games consoles and pretty much anything else with a screen. If a diligent student today fancied recreating the Hamilton walk along the Royal Canal in virtual reality, a VR headset could help us zoom from one side of that now world-famous bridge to the other – perhaps to get a better look at a rare chiffchaff flying overhead. If our computer whiz-kid wants to keep the VR image sharp even during sudden movements, they'll need to encode the digital rotations in terms of quaternions, since the mathematics of quaternions is an essential part of the fastest image-processing algorithms. Closer to home, matrices are indispensable for modern information processing. Google's renowned PageRank algorithm, for instance, works by uncovering the eigenfrequencies of a massive matrix. Matrices also underpin the algorithms used in hedge funds, power AI innovations like ChatGPT, forecast your sunny weather and calculate all the roads that lead to Rome in Google Maps.

The simplest example of a classical many-particle system is a string (like those on a guitar, piano or mandolin). The mathematical model of a string starts with the idea that it's made up of countless atoms (particles) that influence one another. Take the D-string, for instance: it's described by a D-string matrix, whose eigenfrequencies (the different tones) dictate how the string sounds. When you pluck the D-string, it doesn't just produce the fundamental tone; higher notes also resonate, each one a multiple of the fundamental frequency. Every string has its own unique matrix and eigenfrequencies, which means it can only produce a limited, discrete set of tones. In physics, this 'limitation' is called *quantization*. When you strike a string, all these allowed tones resonate at once, though some more prominently than others. And that leads us neatly to our next crucial concept: *superposition*.

In the illustration below, an unwitting hand plucks the D-string. Points A and B mark where the string is attached. The sound produced is a sum (or superposition) of various tones, determined by the shape (or function) of the string at the exact moment it is plucked. This fascinating property, where any function can be expressed as a superposition of different waves, forms the foundation of Fourier analysis, one of the most powerful tools in mathematics. French mathematician Jean-Baptiste Joseph Fourier (1768–1830) developed Fourier analysis to describe how heat moves through matter. Ironically, he struggled to regulate his own body heat. Perpetually cold, he layered himself in sweaters – yet still shivered. In the final days of his life, Fourier resorted to living in a black cardboard box, the only place where he found warmth and comfort.

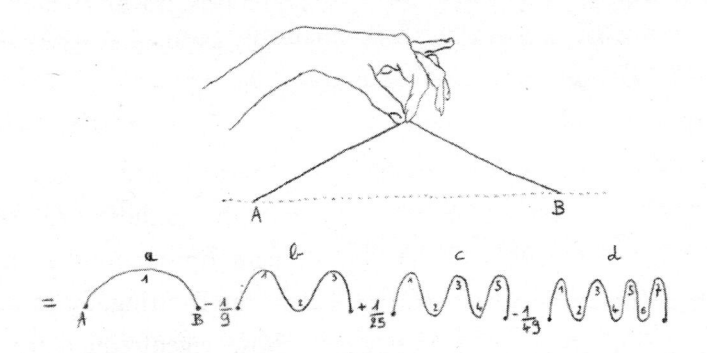

The triangular shape of the string is a superposition (sum) of a fundamental tone (a) and various overtones (b, c, d, etc.), all of which are multiples of the fundamental.

The timbre of an instrument is shaped by the volume of its overtones (how much the string oscillates up and down) and the exact moments when these vibrations hit their peak (known as the phase). In theory, any combination (or superposition) of amplitudes and phases of these tones is possible. For example, the fundamental frequency in the figure is 1. Overtone *b* has a frequency of 3 but is nine times quieter than the fundamental. Overtone c has a frequency of 5 and is twenty-five times quieter, while overtone d has a frequency of 7 and is forty-nine times quieter than the fundamental.

This might feel like diving into technical territory a bit early but bear with us – it's worth it. Here's why: one of the most important breakthroughs by quantum pioneers Heisenberg and Schrödinger was the realization that the behaviour of a single quantum particle (like an electron) follows the exact same mathematical principles that describe many classical particles (like a vibrating string). Sounds odd, right? Defining a single drop of water by studying an entire wave? How does that work? Simple: by turning your intuition off and quantum logic on. Once you do, it becomes clear: the possible energy levels of a quantum particle are *quantized* (hence the name quantum physics). A quantum particle exhibits both particle-like and wave-like behaviour. Like a string, it exists as a superposition of waves, which means it can also occupy multiple positions simultaneously.

In a nutshell, quantum particles are governed by wave equations – what we call the *wave functions* of quantum physics. The evolution of these wave functions is dictated by the Hamiltonian (named after Sir William Rowan Hamilton), an infinitely large matrix that forms the foundation of all quantum calculations, and its eigenfrequencies. If this all sounds a bit abstract or overwhelming, don't worry. In the coming chapters, we'll ride these waves together and unravel these concepts in greater detail.

IN A NUTSHELL

> Symmetry is the most powerful concept in physics. The laws of nature only know one god: symmetry.

> Symmetry: break it to make it! In nature, all structure and order emerge from the breaking of symmetry.

> Group theory is the language of symmetries.

> *Dramatis personae*: Emmy Noether, Lev Landau, Évariste Galois, Wolfgang Pauli.

TWO

SYMMETRY

2.1 The order of the chaos

The *grande dame* of symmetries in physics is the incomparable Emmy Noether (1882–1935). Upon her untimely passing, Einstein shared these words in a letter to *The New York Times*:

> *In the judgment of the most competent living mathematicians, Fräulein Noether was the most significant creative mathematical genius thus far produced since the higher education of women began. In the realm of algebra, in which the most gifted mathematicians have been busy for centuries, she discovered methods which have proved of enormous importance in the development of the present-day younger generation of mathematicians.*

Noether's brilliance lies in her realization that the laws of physics governing change and motion also conceal a fundamental constancy. She uncovered the key to understanding why the laws of physics are the way they are. While Newton and Galilei displayed genius by 'guessing' specific laws, Noether looked deeper and discovered an organizing principle behind them all: symmetry. She demonstrated that Newton's conservation laws are not isolated truths but direct consequences of these symmetries. Symmetries, as it turns out, are the building blocks of nature. They're woven into the very DNA of physics, and their most exquisite application is found in quantum mechanics.

Emmy Noether

Here's the (slightly technical) reasoning behind her theorem: when an apple falls from a tree (and promptly gets eaten), its total energy doesn't disappear – it's conserved. This conservation of energy stems from the fact that the laws of nature don't change over time. Similarly, momentum (the 'amount' of motion) is conserved because these laws hold true everywhere. These principles are, respectively, time-invariance and translation-invariance. But there's more. Kinetic energy is proportional to the square of velocity. This relationship arises because the laws of physics are symmetric (invariant) under Galilean transformations. These transformations, named after Galilei's famous thought experiment, propose that you wouldn't notice any difference between experiments conducted on a ship in motion and those conducted on one at anchor. Einstein's theory of relativity is also all about symmetries. It builds on and expands Galilean transformations, ensuring that interactions don't happen instantly but at the speed of light.

THE NOETHER BOYS

As a student, Emmy Noether often found herself relegated to the back of lecture halls – not by choice, but because women in Germany were still barred from pursuing higher education at the time. Eventually, however, she was welcomed into the mathematics faculty of the University of Göttingen by none other than David Hilbert, widely regarded as the greatest mathematician of his age. When critics objected, Hilbert famously shot back: 'I do not see that the sex of the candidate is an argument against her admission. We are a university, not a bath house.' Once admitted, Noether quickly earned respect and admiration. Her students, fondly dubbed the 'Noether boys', became loyal disciples, while she rose to prominence at lightning speed. With her unrelenting enthusiasm, she didn't just teach; she revolutionized mathematics itself. Noether laid the foundations of category theory, a new framework for mathematical thinking that remains one of the most dynamic fields in mathematics and theoretical physics today. By the way, it was Noether who proved that energy is conserved in the theory of general relativity, a problem that had captivated the brightest minds in science, including those of Hilbert and Einstein.

In 1933, as the Nazi regime tightened its grip, Noether was forced to flee Germany. She found refuge in the United States, where she was warmly welcomed as a professor at Bryn Mawr College in Pennsylvania, an institution that prided itself on empowering women. There, the benches and lecture halls were filled not with 'Noether boys', but with 'Noether girls'. Tragically, her time at Bryn Mawr was cut short. Just two years later, at the age of fifty-three, Noether died following surgery to remove a tumour. It was a life that ended far too soon, but her legacy continues to shape mathematics and physics to this day.

David Hilbert

Nearly all major breakthroughs in theoretical physics stem from discovering new ways in which symmetries manifest themselves in nature. The simplest example is Mendeleev's periodic table, where the periodicity reflects how the rotational symmetry of an atom is represented. A more advanced example of symmetry shaping the laws of nature is the Standard Model, which defines the properties of all elementary particles. This model is entirely built around a symmetry group. In essence, quantum physics is a blend of three vital ingredients: a pinch of symmetry, a pinch of conservation laws and a pinch of evolving insight. The perfect recipe for a three-layer cake. And once again, we're faced with the almost *unreasonable effectiveness* of mathematics: one key can unlock an entire series of doors. In this sense, Emmy Noether held a true master key in her hands.

Behind every strong woman is a strong man. The *grand homme* of symmetries in physics is the much-lauded Lev Landau (1908–1968). Where Noether's theory demonstrated that the laws of physics are dictated by symmetries, Landau, building on ideas from Pierre Curie (who will make an appearance later), discovered the key to bringing order to the rich complexity of matter. And you get that structure, curiously enough, by looking at the way symmetry *breaks*. It's when symmetry shatters that things truly get fascinating. Landau preferred to pick up the pieces.

THE BIG LANDAU SHOW

Several decades after his death, Lev Landau was brought back to life – at least in spirit – in a colossal and eccentric project called *DAU*. This ambitious spectacle, a bizarre fusion of *Big Brother* and *The Truman Show*, ran from 2008 to 2011 and was funded by a Russian oligarch. For three years, 400 characters and 10,000 extras lived in complete isolation from the modern world in a meticulously recreated Stalinist environment. Everything – down to the roubles, food, underwear, cigarettes and even the social rules – was faithfully restored to mirror life in the Soviet Union from the 1930s to the 1950s. Participants lived, worked, ate and slept on set. The commitment was so thorough that, over the course of this shared existence, no fewer than fourteen children were conceived. At the centre of this extravagant production was none other than Lev Landau himself, played by Greek–Russian conductor Teodor Currentzis. The story unfolded in and around a faithfully reproduced version of the secret research institute where Landau worked from the late 1930s until his death in 1968. The illusion of this parallel world was maintained at all costs. In fact, older participants who had lived through the Soviet era and tried to smuggle genuine roubles onto the set were charged with fraud. The result: 700 hours of footage – shot by a single cameraman – that formed the basis for a dozen films, a documentary, a series and more. What did we ultimately learn about Landau? Not much, truth be told. *DAU* was more about the complexities of identity: what it means to be yourself when you can't fully *be* yourself in a constructed environment that only works if you play along as your authentic self. Or something like that. What *DAU* does make clear, though, is just how much Lev Landau continues to capture the imagination.

2.2 When symmetry breaks

Emmy Noether's house provides the perfect setting for a brief introduction to symmetry and symmetry breaking. In her living room there's a square coffee table with a goldfish swimming in a round bowl perched on top. Above the table, a canary in a glass cage dangles gracefully from the ceiling.

Let's dive into the world as seen through the goldfish's eyes – two microscopic lenses gazing at the world like a seasoned physicist. The fact that the water isn't an infinite mass (since the bowl is neatly confined by a glass boundary) is not important here. When we peer through the goldfish's microscopic eyes, we notice that, at the molecular level, everything looks the same. Why? Because the positions of the water molecules are so randomly arranged that they appear identical no matter the angle from which you look. The goldfish can twirl around, roll its eyes or do pirouettes to its heart's content – the view of the water molecules will stay unchanged. In this sense, water shares a property with the circular opening of the fishbowl: no matter how much you rotate the bowl (or the circle) around its axis, it will always look perfectly round, or: infinitely symmetric. Gases share this property as well.

Being meticulous as ever, Emmy has placed the fishbowl precisely at the centre of her perfectly square coffee table. If she rotates the table 90 degrees around its axis, the canary perched above sees no change in the table, or the bowl. The same holds for rotations of 180 degrees, 270 degrees or a full 360 degrees. Each 90-degree rotation returns the corners to their original positions. But if Emmy were to rotate the table by, say, 137 degrees, 45 degrees or 314 degrees, the canary will notice something: while the fishbowl still looks the same, the four corners of the table have shifted. The conclusion is clear: a square has less symmetry than a circle.

Let's take this one step further. Our ultimate goal is to focus on symmetry at the molecular level, but a little living-room physics helps set the stage. Suppose the temperature in Emmy's living room drops below freezing. The water in the fishbowl solidifies into ice. The table and the fishbowl remain where they are, and the goldfish will not look any different either. But what the goldfish perceives will have changed: the molecular structure of ice is different from that of liquid water. The symmetry of ice is comparable to that of the table: if the ice crystal isn't rotated in a specific, well-defined way, its structure looks entirely different. In other words, ice has less symmetry than water.

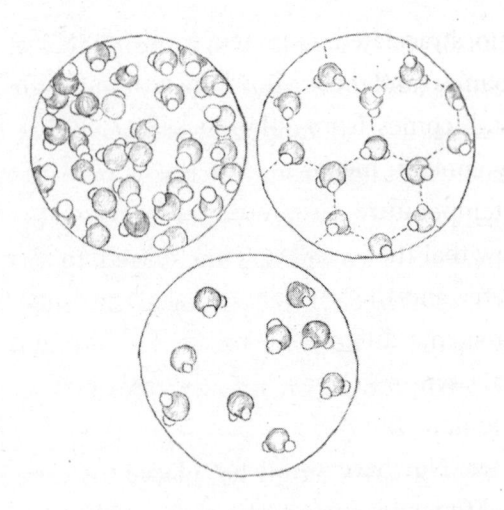

H$_2$O molecules in water (top left), ice (top right) and gas (bottom).

Many a reader is likely scratching their head right now (we get it). The idea that water (top left) is more symmetric than ice (top right) sounds as counterintuitive as saying that ice is lighter than water. Especially when you look at this illustration. But that's exactly the paradox – and it explains why scientists struggled for so long to grasp this subtlety. Even if the theory doesn't seem entirely watertight, it demonstrates one thing above all: human intuition is a terrible guide in physics. To understand why water is more symmetric than ice, you need to abstract the concept of 'symmetry' mathematically. And the quantum cabinet of curiosities has even more surprises in store. Because water and gas share the same symmetry, they can transition seamlessly into each other, without a sound, without a scent and without a phase transition (as long as the pressure is high enough). The key point is this: when water cools to below freezing, symmetry breaks. At least, it breaks a little bit. There's still a lot of symmetry left. Depending on the pressure and temperature, you end up with one of nineteen possible types of ice crystals, each with its own unique symmetry and, if you will, its own 'flavour': strawberry, vanilla, pistachio, stracciatella. In science, of course, they have much less appealing names (and they're nowhere near as tasty). Incidentally, the word 'crystal' comes from the Greek *krustallos* (κρυσταλλος), which, fittingly enough, means . . . ice.

When the temperature rises and the ice melts, symmetry is restored. So now that the ice is broken, we can conclude: the colder, the less symmetry and the warmer, the more symmetry.

The realization that different phases of a material can be distinguished by their symmetry marked a major breakthrough. With this insight, Landau discovered a master key to characterize all phases of matter. Nowhere more than in quantum physics does symmetry breaking take on a wide range of diverse and subtle forms. Many quantum systems owe their most intriguing properties not to their symmetry, but to the specific ways in which that symmetry is broken.

MAGNETIC MISFITS

A concrete example of symmetry breaking is magnetism. The relatively unsung Hendrika Johanna van Leeuwen (1887–1984) tackled the problem of symmetry breaking in magnets (the magnetic field of a magnet always points in one specific direction). Her study of spinning charges led her to a striking conclusion: according to the classical physics of Newton and Hamilton, magnets shouldn't exist at all. Magnetism, she realized, must be a purely quantum mechanical phenomenon, one that inherently involves symmetry breaking. Niels Bohr, soon to become a central figure in this narrative, independently arrived at the same insight. Together, their discovery became known as the Bohr–van Leeuwen theorem.

If you're wondering how symmetry breaking can be reconciled with the fact that the laws of nature are symmetric everywhere: it can't. That's precisely the point, it doesn't add up. Symmetry breaking is an example of emergent behaviour, which only occurs in systems with many particles. Emergence means that in a system of many particles, entirely new laws apply that don't hold for individual particles. One lone starling in the sky won't turn any heads, but a whole flock in motion forms mesmerizing, ever-shifting patterns that defy simple explanation. In short: $1 + 1 \neq 2$. That's a constant in the quantum story. On this point, Aristotle was right: the whole is much greater than the sum of its parts.

A less ethereal but thoroughly quantum example of symmetry breaking (and by extension, emergence) is superconductivity (see Chapter 8). Superconductors have the remarkable property that, when paired with a magnetic field, they cause objects above them to 'float'. Why? Because symmetry breaking literally expels the magnetic field from the superconducting material. Japan's high-speed Maglev trains, which hover above their rails at speeds of up to 375 miles (more than 600 km) per hour, are a real-world application of this principle. And as for Maglev, it's no coincidence that the name nods to 'Magnificent Lev Landau'!

2.3 Groups: the structure behind symmetries

Just as we've elevated Simon Stevin to the ranks of personal nobility, it would be a disservice to the mathematical revolution not to place Évariste Galois (1811–1832) on a well-deserved pedestal. His story, like that of many others to follow, shows how the insights of a single person into one specific problem can spark a true revolution. Yet, Galois's theories were so intricate that they were often dismissed as 'too crazy to be true'. Hamilton's breakthroughs might still have been digestible, but Galois's . . . his were a bridge too far.

Galois sank his teeth into the roots of polynomial equations of the fifth degree and beyond, a problem that had eluded mathematicians since Babylonian times. Sounds complicated? It is. In 1824, the Norwegian mathematician Niels Abel (1802–1829) had already delivered the liberating verdict: there is no solution, sorry. Yet, there actually is an answer: it is not solvable. Case closed? Not for Galois. Reluctant to let the matter rest, he spotted a pattern in Abel's work. He constructed functions based on the solutions (roots) of polynomial equations, revealing a remarkable symmetry: permutation symmetry (where the functions remain unchanged even when the order of the roots is switched). He also discovered that for fifth-degree equations, this permutation symmetry is far more complex than for lower-degree equations.

Galois broke open the rigid confines of geometry, invented group theory and became the first to wield the incredible power of symmetry in mathematics. *Show me your symmetry, and I'll tell you who you are.* Group theory, he proved, could unlock some of the most fundamental problems in mathematics. And their influence didn't stop there. In quantum physics, too, everything from electrons and neutrons to bosons and quarks can be classified and understood through their symmetries. Just as Newton created a mathematical framework – calculus – to describe accelerating objects, Galois developed the mathematics for describing symmetries. Consider this: without group theory, whether in the form of permutation groups, matrices, transformation groups or other abstract algebraic constructs, the modern natural sciences as we know them wouldn't exist.

A CRASH COURSE IN GROUP THEORY

1. A group is a set of operations performed on one or more elements while preserving certain properties. For example: if we rotate Emmy Noether's square coffee table (the element) 90 degrees around its axis (the operation), the table's start and end positions are identical. In contrast, for a circle (or a fishbowl), you can perform an infinite number of rotations without changing its appearance.
2. The second property of a group operation is that the combination of two operations can always be reduced to a single operation. For instance, rotating the table 90 degrees twice has the same effect as a single 180-degree rotation. This is known as group multiplication, and can be expressed as: $g_1 \cdot g_2 = g_{12}$.
3. The third property is where things get interesting: the most important groups in quantum physics are *non-commutative*. In a group, $(g_1 \cdot g_2)$ is not necessarily equal to $(g_2 \cdot g_1)$. For example: rotating a triangle and then reflecting it gives a different result than reflecting it first and then rotating it. This non-commutativity is precisely what makes quantum physics so elusive. It also underpins the famous principle in quantum physics that we cannot simultaneously know both the position and velocity of a particle – with all the spooky consequences that follow.
4. Finally, while group operations don't necessarily have to be commutative, they must always be associative. Associativity is the defining – though far from intuitive – fourth property of groups. It means that the way operations are grouped

> doesn't matter, as long as the sequence of the individual elements (here g_1, g_2, g_3) remains unchanged. In other words: $(g_1 \cdot g_2) \cdot g_3 = g_1 \cdot (g_2 \cdot g_3)$.

Since a group multiplication is always associative, mathematics dictates that individual group elements can be represented as matrices (thank you, Sir Hamilton!). These very same matrices form the mathematical foundation of quantum physics. This is why group theory is so essential. The matrices from group theory provide a concrete way to describe how wave functions transform under symmetry operations. Wave functions, represented by the symbol ψ (psi), contain all the information about the system under study. From them, you can determine the probability of particles being in certain locations, how they interact and how they evolve over time. If you understand how a wave function transforms under a symmetry operation, you can also identify which wave functions remain unchanged (invariant) under that transformation and how to represent them. This principle is at the core of quantum physics' success. The clearest example of this success is Mendeleev's periodic table, where all atoms are arranged according to their symmetry structure. To understand the structure of the table, you only need to analyse all possible wave functions that remain invariant under rotations and reflections – the only symmetries of a (spherical) atom. But that's a topic for Chapter 6.

Évariste Galois's true passion lay in permutation groups – the set of all possible ways objects or particles can switch places. These groups are central to quantum physics because, unlike in classical physics, atoms and elementary particles are indistinguishable. Their properties remain unchanged under permutations. In quantum physics, you can't simply say: this is particle 1 and this is particle 2. If you calculate the probability of particle 1 being here and particle 2 being elsewhere, the result is identical when the particles are swapped. This indistinguishability isn't just a curious quirk – it's the very reason all particles in the universe fall into just two categories: bosons and fermions.

The most well-known bosons are photons, the particles that make up light. Thanks to the permutation symmetry of their wave function, an infinite number of bosons can occupy the same state. This property enables us to see light, create lasers and build Bose–Einstein condensates (thanks to Bose, Einstein and some very cool physics).

The most well-known fermions are electrons. Their indistinguishability is manifested through the anti-symmetry of their wave function, with dramatic consequences: it prevents them from occupying the same state. As a result, electrons repel each other with heart and soul. This led Wolfgang Pauli (1900–1958) to formulate one of quantum physics' most crucial principles: the exclusion principle. It encapsulates by far the strongest force in nature. We owe it to the symmetry property of the electrons that the world doesn't implode, that matter is solid and repels other matter, and that we don't find ourselves swallowed by a tree when hugging it. Without the exclusion principle, the Earth would shrink to the size of a pea and you could quite literally sink through the ground in shame.

Wolfgang Pauli

NOT EVEN WRONG

Fortunately, Herr Pauli didn't drop by for coffee during our goldfish experiment at Emmy Noether's house, otherwise the poor creature would have drowned on the spot. Pauli was infamous not only for his razor-sharp wit but also for the strangely destructive effect his mere presence had on technical equipment. If he happened to be near an experiment, that experiment was almost guaranteed to go awry. This peculiar phenomenon became known as the *Pauli effect*. On one occasion, a group of physicists in Milan decided to put the legend to the test. They rigged up a device designed to emit a loud and perfectly timed *bang* the moment Pauli opened the door. They tested it repeatedly to ensure it worked (and it did). The ultimate twist? When Pauli finally entered the room, nothing happened at all. The Pauli effect had, quite brilliantly, sabotaged itself.

Despite his tendency to dismiss nearly every question as nonsensical, Pauli was endlessly generous and patient with colleagues and students. That said, he had little tolerance for

> sloppy thinking and was merciless in dismantling arguments that didn't hold up, delivering his now-legendary verdict: 'That's not just not right. It's not even wrong!' (*'Das ist nicht nur nicht richtig, es ist nicht einmal falsch!'*). With his trademark *not even wrong*, it was abundantly clear who ruled the roost in the world of quantum physics at the time.

2.4 Drums and atoms

The (abstract) concept of group theory is best illustrated with the vibration of a drumhead. Think of a drum as a two-dimensional version of a (one-dimensional) string. The key difference is that a string has only reflectional symmetry, while a drumhead has both reflectional and rotational symmetry. However, these two types of symmetry are non-commutative: reflecting first and then rotating is not the same as rotating first and then reflecting. This non-commutativity affects the fundamental vibrations, some of which will break the symmetry and therefore occur in pairs, each with the same energy and frequency.

Here's a thought experiment: place a drum next to a sturdy set of speakers, sprinkle a handful of (uncooked!) rice grains onto the drumhead and blast a few high-pitched notes through the speakers. Just like the string in Chapter 1, the drumhead exhibits quantization – it will only resonate at specific, well-defined frequencies. This resonance becomes beautifully visible in the distinct (wave) patterns that emerge in the rice on the drumhead, known as Chladni patterns.

SOPHIE GERMAIN

The French mathematician Sophie Germain (1776–1831) was the first to muster the courage to tackle the mathematical explanation behind Chladni patterns formed by vibrations on a drumhead. Even the most renowned mathematicians of her time shied away from the problem due to its daunting complexity. But Germain embraced the challenge. Her conclusion: the lines where the rice collects correspond to points where the wave displacement is zero. These wave patterns can be determined through a combination of calculus and group theory.

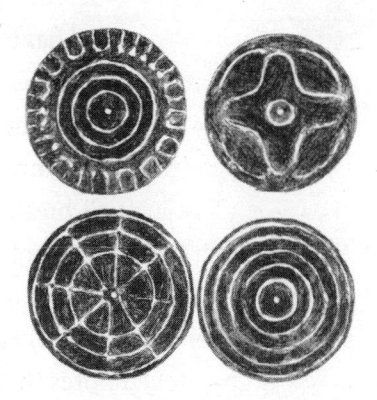

Chladni patterns: the white lines are the places where the rice gathers (the nodes). Here, the wave doesn't move.

The images on the next page help make this concept more concrete. Each circle represents a drumhead viewed from above. In the black areas, the drumhead moves upwards, while simultaneously moving downwards in the white areas. Then the process reverses: the white areas rise as the black areas descend. This oscillation occurs at a specific, well-defined frequency.

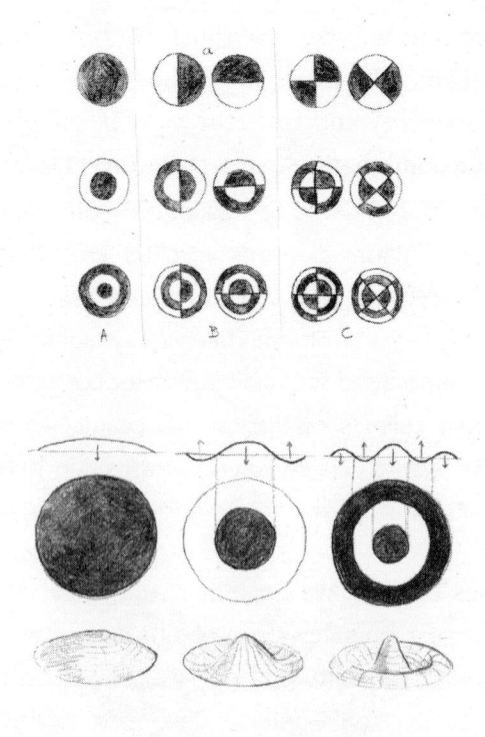

Above: the fundamental vibration patterns of a drum; every other vibration is a superposition of one of these vibrations.
Below: close-up of column A: the vibration of a drumhead, analogous to the perfectly symmetric S orbitals.

In the left column (A), the patterns are circular, making them perfectly symmetric. They exhibit the same rotational symmetry as the circular drumhead or the evenly spreading ripples that form when you drop a pebble into water.

The symmetry of the waves in the middle column (B) and the right column (C) is more complex. These waves are 'degenerate': they occur in pairs. Each pair represents the same type of wave, but vibrating in different directions. Take the two vibrations in the top illustration of the middle column (a), for instance: by applying

52   *Why Nobody Understands Quantum Physics*

a symmetry operation, you can transform one vibration into the other. Apart from a 90-degree rotation, they are identical; they share the same energy and frequency. Every possible vibration of a drumhead is a combination (superposition) of these fundamental vibrations.

This illustration shows a little more clearly what the superposition of two (fundamental) vibrations can look like, although we could have equally rotated them in the other direction.

In the previous chapter, we explained that in a one-dimensional world, the state of a single electron can be described as a superposition of the fundamental vibrations of a string. In a two-dimensional world, this same electron would be described by the superposition of the fundamental vibrations of a drumhead. Real atoms, however, with electrons orbiting their nucleus, exist in a three-dimensional world. This means we need to reimagine our two-dimensional drumhead as a three-dimensional drumhead.

It stands to reason that the symmetries of a three-dimensional drumhead (reflections and rotations in 3D space) are far more intricate than those in two dimensions. This is largely because they are inherently non-commutative. In this case, the fundamental vibrations are represented in a highly specific way. Some abstraction is required here – after all, thinking in multiple dimensions is no small feat! The illustration overleaf displays all the possible fundamental vibrations of a 3D drum, analogous to the earlier 2D drumhead illustrations. These fundamental vibrations are labelled S, P, D and F. The S vibrations are perfectly spherical and fully symmetric. The P vibrations exhibit three-fold degeneracy and resemble a roller coaster that can oscillate in three distinct directions (any other P vibration is simply a superposition of these three fundamental P vibrations). The D and F vibrations are even more complex, with five-fold and seven-fold degeneracy, respectively.

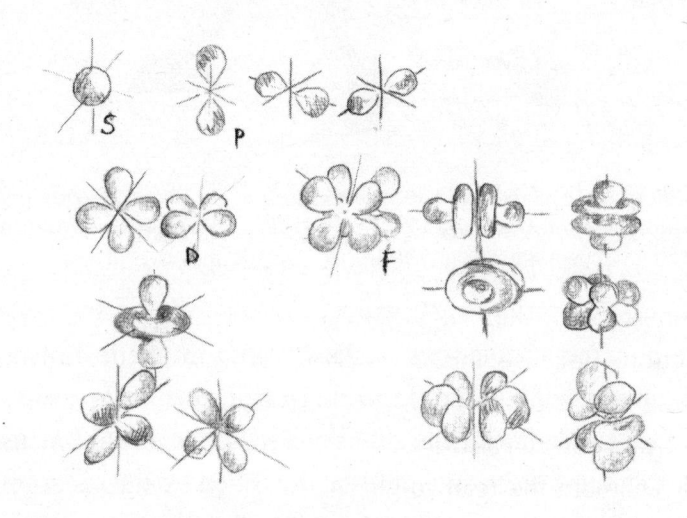

Electrons move around the nucleus in orbitals, or 'clouds'. The S clouds are perfectly round, making them fully symmetric, and can hold up to two electrons. The P clouds are a sum of three 'sub-clouds', each capable of holding two (of six) electrons. These sub-clouds extend through the atomic nucleus, forming a characteristic dumbbell shape. The D and F clouds are even more complex, splitting into five and seven 'sub-clouds' respectively, each accommodating two electrons.

In the realm of quantum physics, this S, P, D and F classification forms the foundation of our understanding of the electronic structure of atoms. The fundamental vibrations correspond to the energy orbitals where electrons can reside and, together with Pauli's exclusion principle, they explain the entire structure of Mendeleev's periodic table. As phenomenally diverse, original or danceable as drum riffs may be, they pale in comparison to the vastly more intricate and expansive three-dimensional world of atoms.

IN THE CLOUDS OF ELECTRONS

We kindly invite everyone to abandon the notion that electrons move in simple energy orbitals around the atomic nucleus, like planets orbiting the sun. Instead, think of an electron as something highly elusive, with no fixed shape or trajectory. It exists everywhere and nowhere simultaneously around the nucleus, like a wisp of mist or a cloud. This cloud, in turn, is a sum (superposition) of various fundamental vibrations of the electron. Because this concept is so abstract, we use the simpler terms orbitals or shells to describe it. In this context, superposition can also be understood as 'supposition': we can only estimate the relative probability of an electron being in one place or another. It's always a matter of educated guesswork with electrons. But one thing is certain: they exist.

P orbitals are elusive, like drifting wisps of mist.

PART TWO

QUANTUM

After Stevin's resounding *thud*, one mathematical rule after another was discovered, shaping the grammar of the language in which nature's book is written. Mathematics gained its rules, physics gained its mathematics and humanity gained its physics.

Then, suddenly, the whirlwind of the twentieth century arrived. Hot water had been invented, the skies had been mapped and society stood firmly on its foundations. With the formulas and universal laws at hand, almost everything seemed explainable. If something could be studied, weighed or measured, it was considered understood. And whatever lay beyond the reach of science was conveniently relegated to religion. Since Antonie van Leeuwenhoek (1632–1723) had discovered a teeming world of 'very small creatures', it seemed as though nature had revealed all her secrets. Or so they thought. But the best was yet to come. As the new century dawned, a genuine revolution unfolded through the entire cultural and intellectual landscape. Society was ready for something new. It was time for modernity. From then on, everything was conceived in terms of 'bigger' and 'more'.

In science, however, the focus shifted to increasingly smaller scales. This inevitably raised new questions and cast doubt on existing knowledge. Certainties began to falter, clarity blurred and intuition faced unprecedented challenges. Why is matter solid? Why doesn't the sun burn out? Why is grass green? Why do hot objects change colour? Why are things the way they are? Something peculiar was going on with atoms, electrons and all those tiniest of particles, something beyond the reach of existing formulas and theories. Both theory and experiment found themselves at a bit of a loss. Had something fundamental been overlooked?

The irreversible nature of perception: as soon as we understand or see something, it's sometimes impossible to imagine that we didn't always understand or see certain things. It's like the superposition of the nude descending a staircase: once you have seen it, you can no longer unsee it. À la Marcel Duchamps, Nu descendant un escalier.

The following chapters illustrate how quantum physics emerged thanks to giants standing on one another's shoulders, each seeing just a little further. Newton himself openly acknowledged that much of his work was built on the tireless efforts of those who came before him. But he distinguished himself by going further, thinking deeper and taking more radical leaps. Sometimes, scientists arrived at the same inevitable conclusions almost simultaneously, though independently. Newton, for instance, was convinced that light was made of particles. Shortly after, Christiaan Huygens argued the opposite: light, he claimed, was a wave. Then Einstein complicated matters: light is both particles *and* waves. Enter Prince de Broglie, who upended everything once more. 'It depends on how you look at it,' Heisenberg concluded. Everyone was a little bit right. And inevitably, a little bit wrong. So, in the end, no one was (not) right.

PARALLEL ARTS

During the first quarter of the twentieth century, a revolutionary movement unfolded, not only in science. For artists too, it no longer seemed possible to understand the world in simple terms. The figurative representation of reality was abandoned in favour of more abstract art forms such as cubism, surrealism and expressionism. Much like the pioneers of quantum theory, the creative minds of the time believed in an underlying reality in which everything incompatible came together. They distanced themselves from reality, questioned identity and, from this radical idealism, they created space for imagination and chaos. This shift gave rise to a completely different concept of 'harmony'.

Egon Schiele painted alienating (self-)portraits, Pablo Picasso depicted a woman (Dora Maar) whose face simultaneously gazes outward and inward, and Marcel Duchamp no longer limited himself to one pose, but depicted the entire movement as a whole, portraying a nude figure (a woman?) descending the stairs (or is it a man walking up the stairs?). In his novel *One, No One, and One Hundred Thousand*, Luigi Pirandello's protagonist has an identity crisis and breaks down into a hundred thousand selves. Arnold Schoenberg redefined harmony in music, developing atonal compositions characterized by dodecaphony. This musical form, incidentally, shares striking similarities with quantum mechanics: per sequence, all twelve tones must be used, in whatever order, but only once. The sounds are represented, as in matrix mechanics, by numbers referring to their respective colour, duration and volume and, in terms of symmetry, a sequence of tones can also be reversed. In architecture, art nouveau emerged, frills were exchanged for functionality, beauty once again stood for simplicity – just as in mathematics, which has to not only be correct, but also beautiful. Le Corbusier didn't just design buildings; he reshaped entire spaces and manipulated light to his will. Across all these disciplines, one striking feature stands out: the observer (audience, reader, viewer, listener) is given a very active role and becomes increasingly participatory in 'the work of art'. The real meaning of a work lies somewhere between the original intention of its creator and the final interpretation by the observer.

IN A NUTSHELL

> Max Planck: Packing light – in quanta.

> Albert Einstein: Let's be more general – light is made of particles!

> Ernest Rutherford and Niels Bohr: Forget about the plum pudding. A tiny nucleus, please, with electrons orbiting like planets!

> Louis de Broglie: Think again – electrons are waves. *Non*, everything is a wave!

> Werner Heisenberg: Observing is disturbing.

> *Dramatis personae*: Max Planck, Albert Einstein, Niels Bohr, Ernest Rutherford, Louis de Broglie, Werner Heisenberg.

THREE

THE (IM)PROBABILITY OF A PARTICLE

3.1 Max Planck's great tiny quantum leap

Max Planck: 'Science enhances the moral values of life, because it furthers a love of truth and reverence – love of truth displaying itself in the constant endeavour to arrive at a more exact knowledge of the world of mind and matter around us, and reverence, because every advance in knowledge brings us face to face with the mystery of our own being.'

Physics has a way of driving even the most measured minds to bold speculation. Max Planck (1858–1947), the very embodiment of modesty, grace and politeness, experienced this first hand. The year was 1900. Many had advised the young Planck against pursuing physics – after all, wasn't everything already discovered? Why not follow his other great passion: the piano? He certainly had the talent: perfect pitch, a keen sense of order, years of experience in a boys' choir. Plus, he had a flair for composition. But Planck thought differently. 'Just because everything's been discovered doesn't mean there's nothing left to explore,' he reasoned. Following his moral compass – and his heart – he pressed on. He wanted to (and would) do something good for the world and find the laws to which nature adheres. He was fully prepared to reinvent physics from scratch.

Life, however, was not kind to Planck. His first wife, twin daughters and two sons all died before their time and, as if that were not enough, the Second World War deemed it necessary to pound everything he owned into rubble. Yet, he managed to convert all these calamities into positive energy. To a grieving world, Planck gave revolutionary theories that propelled science into a new era. He also left an enduring legacy in the Max Planck Institutes, which revitalized Germany's place on the global scientific stage after the war.

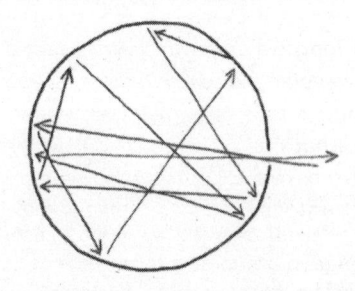

A black box absorbs light of every possible frequency, irrespective of the material the box is made of. Light entering the box is endlessly absorbed by the walls and then re-emitted at various frequencies. As a result, a black box contains every possible frequency. Planck's question was: how much light of each frequency is present?

Let's start (again) at the very beginning. It all began with nothing – or rather, with an empty black box. Until the turn of the twentieth century, no one had managed to explain the behaviour of the electromagnetic radiation emitted by a hollow black box (or 'black-body emitter'). Here's an important side note: every physical object continuously emits electromagnetic radiation from within, because the atoms and molecules that make up matter are constantly vibrating. In fact, 65 per cent of the heat lost by the human body escapes as electromagnetic radiation. Take iron as an example. When heated, the metal changes colour. This happens because the underlying molecules vibrate ever faster as the temperature rises, emitting light of ever higher frequencies. If the colour is red, it means the iron is already very hot. Heat it further and the red shifts to blue.

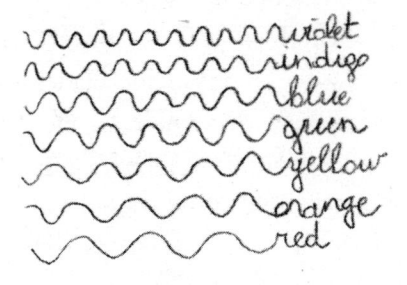

Red light, with its long wavelength, carries low energy. Violet light, with its shorter wavelength, carries more energy. Beyond violet, ultraviolet light and X-rays emerge – both brimming with energy, which is precisely what makes them so dangerous. At the opposite end of the spectrum, red is preceded by infrared rays, which are not visible to the naked eye.

The colour of radiation from a box is determined by the fact that every material absorbs light of specific frequencies (colours) while reflecting others. The grass is greener on the other side because the atoms in your neighbour's lawn absorb only light of frequencies that look slightly less green than those in your own grass. As a result, the neighbour's grass reflects a richer, more vibrant green. That's how it works: what we see is the light (or colour) that is *not* absorbed and is reflected back. We go into more colourful detail on this in Chapter 6.

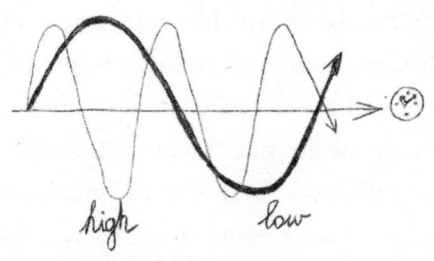

The properties of light are determined by its wavelength (the number of undulations per interval), its frequency (the number of vibrations per second) and its intensity (the height of the wave). The longer the wavelength, the smaller the frequency, and vice versa. The product of wavelength and frequency is constant and equal to the speed of light.

What makes a black box so fascinating is that it absorbs radiation at every possible frequency. Once that radiation is trapped inside the box, it is continuously bounced from one energy level to another through collisions with the walls. The properties of the radiation in a black box are constant and universal, whether the box is made of corrugated cardboard or plasticine, and whether it's oval or square. Black is black.

The experiment that had been getting scientific minds overheated for some time, and on which Planck had fixed his attention, started with an empty black box pierced with a small hole. There was something odd about the radiation (light) escaping from that hole. The aim was to investigate how the intensity and colour (i.e. frequency) of the light changed as the temperature in the box increased. The problem? Experiment and theory were in blatant conflict. But there was also a conceptual issue. In the early years of the twentieth century, it was assumed that everything could be broken down *ad infinitum*, not only distances but also objects and forces – and energy too. Light rays and heat rays, it was thought, could have energy of every possible frequency and intensity. Did that mean the black box was essentially a Pandora's box, emitting ultraviolet light, X-rays or even (much worse!) lethal gamma rays? And if it contains every possible frequency, does that imply infinite energy? According to classical theory, the answer was yes. But that's a catastrophe, an ultraviolet catastrophe! It couldn't be true! Rest assured, it wasn't true. But why not?

Deep within, Planck must have heard a tiny voice echoing from a distant past: if the theory is wrong, then have the 'lead balls' to sweep everything off the table and start again. No sooner said than done, albeit with great reluctance and a bagful of scepticism. Planck cut up the black-box mystery into umpteen pieces and strung the solutions back together. With a clear head and a contradictory yet creative feel for mathematical logic, he came to the conclusion that the energy of light could not possibly be an endless, unbroken current, however much that idea went against his religion (i.e. classical physics). After all: *natura non facit saltus*, nature does not make jumps. Light had to be divided up into little bits somehow or other. It could only be absorbed or emitted in multiples of small packets of light, each with a particular amount of energy.

Planck proposed that the energy of a light wave could only exist as a multiple of a specific 'quantum'. But this raised an important question: how much energy does such a 'light packet' contain? Planck hypothesized that the energy is proportional to the frequency of the radiation, multiplied by an invariant (Planck's constant) which, like the speed of light, is constant and universal, and crops up in virtually every equation in quantum physics. This led him to the following formula:

$$E = h \cdot v$$

E = the energy of a light wave with frequency v
h = the natural constant calculated by Planck ($6.62607015 \cdot 10^{-34}$ joule/hertz)
v = the frequency of the light (derived from the Greek letter 'nu')

Planck concluded that the energy of a wave packet with frequency v must be a multiple of an elementary energy unit: either it's h times v, or twice h times v, or three times h times v, and so on. In short, the intensity of light, and therefore its energy, is quantized. With this insight, Planck was able to fully account for the experimental results.

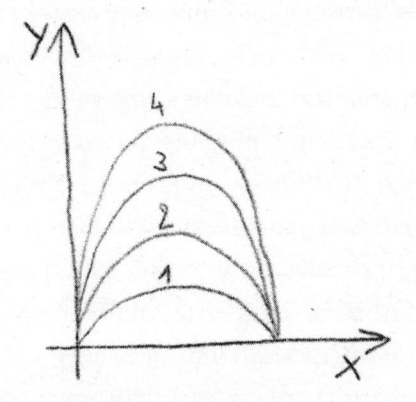

When we pluck a guitar string, the intensity of the vibration can assume every possible value. The same tone can resonate at every possible decibel between 0 and 100. A quantum string is different: its intensity can only assume a few discrete values – the decibels jump in leaps of ten. No intermediate volumes are possible. In a quantum cafe, you might order one bottle of Carlsberg. You can happily order two or three bottles of Carlsberg. But half bottles never cross the counter. That just isn't done.

PLANCK ON THE SHOULDERS OF BOLTZMANN

To arrive at his final formula, Planck relied on several fundamental ideas. As with the string from Chapter 1, all possible frequencies of the light in a box are multiples of a fundamental one. However, it's not just the frequency but also the intensity that is quantized. Here lie the true foundations of quantum physics: light comes in multiples of elementary quanta.

The key question is: how much energy is contained in the black box when the light inside it is 'in equilibrium' (a system is in equilibrium when its properties no longer change over time)? Or more precisely: how many (n) photons of frequency v are present (each with an energy equal to $n \cdot h \cdot v$)?

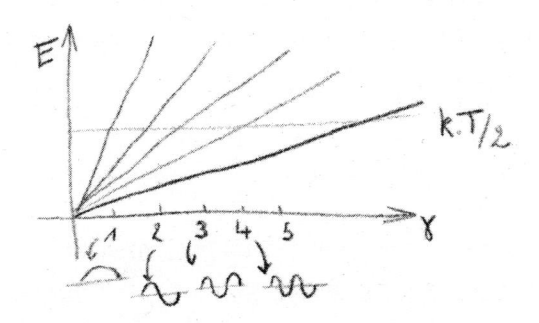

The different straight lines represent the energy of a system of n (n = 1 , 2 , 3 , 4 . . .) photons with frequency v. The height of the horizontal line (labelled k . T / 2) depends on the temperature T. You can determine the number of photons per frequency by identifying where these straight lines cross the horizontal line. x-axis: frequency; y-axis: energy.

THE (IM)PROBABILITY OF A PARTICLE

To arrive at this solution, Planck turned to the statistical physics of Ludwig Boltzmann (1844–1906), which states that each degree of freedom must have equal energy, namely $k \cdot T / 2$ (where $k = 1.380649 \cdot 10^{-23}$ J/K is Boltzmann's constant and T is the temperature of the system expressed in degrees Kelvin).

Because the energy of photons is quantized, Planck had to refine this principle a touch, as multiples of $h \cdot v$ do not fit perfectly into $k \cdot T / 2$. This led Planck to assert that the number of photons n with frequency v is determined such that n is as large as possible but does satisfy $n \cdot h \cdot v \leq k \cdot T / 2$. Suppose that $k \cdot T / (2 \cdot h \cdot v)$ is equal to five, then there are five quanta with frequency v, two with frequency $2 \cdot v$, one with frequency $3 \cdot v$, $4 \cdot v$ and $5 \cdot v$, and zero with all higher frequencies.

This also resolved the problem of the ultraviolet catastrophe. No wave packets with very high frequencies can occur, because even a single photon with a high frequency would have an energy higher than $k \cdot T / 2$.

On the night of his discovery in 1900, Max Planck played Beethoven's 'Ode to Joy' with frenzied intensity. Who said everything had already been discovered? With his hypothesis, Planck had resolved all the issues surrounding black box radiation, and the ultraviolet catastrophe was no longer a threat. Case closed.

QUANTUM IN FIVE WORDS: THERE IS A SMALLEST SCALE

The laws of physics may be universal, but different laws apply depending on the level we are studying. When the coffee machine refuses to work, you're not going to take it apart to see which atoms are to blame – even though the machine is, fundamentally, made up of atoms. The problem exists at a completely different level. As mentioned earlier, Newton's classical theory works perfectly for describing the large-scale (roughly speaking, whatever we can see with the naked eye), but falls short when talking about the very tiniest things. At microscopic level, the symmetry of the classical laws of nature shows some serious cracks.

But what exactly is 'large-scale', and what qualifies as the 'very smallest'? Surely there must be a constant somewhere, something that everything operates in relation to? Could it be another symmetry? No, not this time. And no, it's not a fishbowl either. It's a point at which energy and time, position and speed, start to blur. From that precise point, which is invariable and universal, we need quantum. And that scale is represented by Planck's constant (h).

METRE	OBJECT
100000000000000000	Distance to the second closest star
1000000000000	Diameter of the solar system
1000000000	Diameter of the sun
10000000	Diameter of the Earth
1000000	Distance from Birmingham to the Shetland Islands
100	Height of the Nieuwe Kerk in Delft
1	Height of a tumble dryer
0.01	Diameter of a thimble
0.0001	Thickness of a human hair
0.0000001	Wavelength of visible light
0.000000001	Wavelength of X-rays
0.0000000001	Diameter of a hydrogen atom
0.000000000000001	Diameter of a proton

Orders of magnitude

Planck's heart and head were on a collision course; he couldn't get around the fact that all his calculations had – finally – yielded a theory that tallied with the experiments. That was progress. But it came with a glaring problem: he couldn't explain it. At least not on the basis of classical physics. How could any of it be possible? Stevin could say what he liked, but Planck preferred not to shout from the rooftops and quietly laid his formula aside, prioritizing his principles over glory. It is nothing more than a mathematical trick, he reminded himself, a gimmick, a temporary solution pending a much better explanation. His theory only explained the process of light absorption and emission, yet it still said nothing about the nature of light itself. He (mistakenly) believed that, in the end, light would prove to be a continuous field (i.e. that classical physics did have the right end of the stick). Although . . . his theory seemed to be the only explanation that made any sense.

The experiment with the box proved it in black and white: light consists of quanta.

A QUANTUM FAIRY TALE

Planck was in a quandary. He felt as though he was cheating on the love of his life (the classically beautiful Physica, whom he had come to know through and through) by dallying with Quanta. He had first encountered the breathtakingly recalcitrant Quanta by chance a few months before, and he was utterly unprepared for her boundless, unpredictable energy. Ever since their encounter, he had been in a spin. His world was turned upside down; everything he had held to be true was becoming confused. Meanwhile, the outside world was gossiping about his love for his side affair. 'An empty box!' they scoffed. 'A catastrophe!' But Planck fought back. She made everything fall into place. She bestowed her love in small pieces, discreet as she was. By seeing her in a new light, she began to look completely different. He embraced her, and she became his anchor, the constant in his life. She propelled his energy to higher levels. Together, they were a magical formula, the reconciliation of heart and mind! His heart made a quantum leap. And on 14 December 1900, their baby was born: Quantum. The enchanting newbo(h)rn soon won over his godmother Emmy and the three kings Albert, Niels and Louis, who gratefully showered the child with myriad little packets: photoelectric magic rays, a Lego box full of energy and a golf course of princely proportions. The child's eyes twinkled like stars. With the arrival of baby Quantum, everyone quickly began to relativize and view the world in a fresh, new light.

3.2 Light: wave and particle

Albert Einstein: 'Imagination is more important than knowledge. Knowledge is limited. Imagination encircles the world. [. . .] The mind can proceed only so far upon what it knows and can prove. There comes a point where the mind takes a leap – call it intuition or what you will – and comes out upon a higher plane of knowledge, but can never prove how it got there. All great discoveries have involved such a leap.'

With Max Planck, an unstoppable force was set in motion, shaking the very foundations of all sacred principles. Nevertheless, the black box experiment remained an enigma and no one got much further with the theory that Planck had distilled from all his brainwork.

That is, until 1905, when Albert Einstein (1879–1955) suddenly burst upon the scene. Erm . . . who? Eyes rolled, eyebrows furrowed. Back then, the name Einstein didn't exactly light up the room. Known as *der Depperte* ('the dopey one') since his eccentric youth, Einstein had a penchant for sinking his teeth into warped lines of reasoning that, as it turned out, were beautifully correct in hindsight. Unlike the rather more uptight Planck with his prissy pince-nez, Einstein was indifferent to common belief, consensus or tradition. He owed his 'dopey' nickname to his slightly troubled relationship with language. By some accounts, he didn't speak until he was four, with his very first full sentence – at the lunch table, we presume – being: 'The soup is too hot.' Clearly, his sense of timing (and feeling for time) was served to him from an early age.

Einstein had a secret weapon that came in especially handy where reason fell short, and that was imagination. Which he combined with another powerful tool: the thought experiment. His formulas, which in themselves were not even all that terribly complicated, sprang mainly from a hefty dose of creativity, plus a highly developed form of what-ifism. In 1905, much like Newton in 1665, Einstein experienced his very own *annus mirabilis*. Within the space of four months, he turned the scientific world completely inside out. The tone was set with a letter written in May 1905 to his close friend Conrad Habicht, in which Einstein announced his revolutionary articles:

> *Dear Habicht,*
> *Such a solemn air of silence has descended between us that I almost feel as if I am committing a sacrilege when I break it now with some inconsequential babble . . . So, what are you up to, you frozen whale, you smoked, dried, canned piece of soul . . . why have you still not sent me your dissertation? Don't you know that I am one of the 1½ fellows who would read it with interest and pleasure, you wretched man? I promise you four papers in return, the first of which . . . deals with radiation and the energy properties of light and is very revolutionary, as you will see if you send me your work first.*[1]

Einstein was the first to grasp the revolutionary implications of Planck's theory: what if quantization (where energy is a multiple of

---

1 Walter Isaacson, *Einstein: His Life and Universe*, Simon & Schuster, 2017, p.93.

the smallest quantum) was not merely a mathematical construct but a new and fully fledged physical principle? This wasn't just an elevated tool for explaining phenomena like light processes. What if Planck had uncovered *the* defining property of light? What if this wasn't just a hastily cobbled-together theory that made everything make sense? Einstein boldly and unreservedly threw a cat among the pigeons – and shed a different light on the matter: what if light was both a wave and a particle?

Planck himself didn't fully realize that his theory had laid the groundwork for an entirely new physics. But thanks to Einstein's relentless pursuit, Planck's years of hard work and determination were anything but wasted. Einstein took the momentum Planck had started and supercharged it, stepping even further away from classical theories and daring to think beyond their limits. From his theoretical explanation of the photoelectric effect, Einstein derived the dual nature of light – *both* wave and particle. Interestingly, it was for his explanation of this photoelectric effect, and not for his theory of relativity (also dating from 1905), that Einstein was awarded the Nobel Prize for Physics in 1921, three years after Planck.

THE PHOTOELECTRIC EFFECT

When a source of light is directed at a negatively charged metal plate, electrons are knocked off the surface of the plate and ejected. When the frequency of the light is increased (by using ultraviolet light, for example), the electrons are ejected from their orbitals with more energy. But here's the kicker: as Philipp Lenard (1862–1947) discovered through his experiments in 1903, the energy of these electrons doesn't depend on the intensity of the light. Nor does reducing the intensity necessarily mean it takes longer for electrons to be ejected. It simply results in fewer electrons being released.

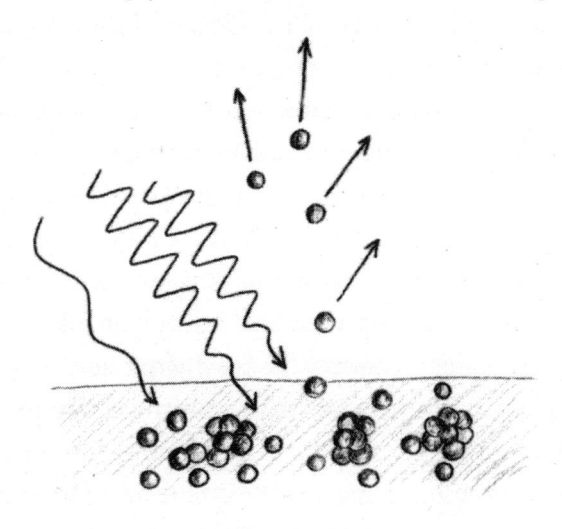

The photoelectric effect: light that hits the surface with a sufficiently high frequency can dislodge electrons from a metal plate.

Much like the black-box phenomenon, the photoelectric effect stood in stark contradiction to classical physics. According to classical theory, low-intensity light should take longer to eject an electron, as the electron would need to gradually absorb enough energy from the light wave. It also claimed that the energy of the electrons would be proportional to the light's intensity. And there was yet another fly in the ointment. Classical physics suggested that electrons should be ejected by light of *any* frequency. Yet Lenard's experiments clearly showed that there's a minimum frequency (completely independent of light intensity) below which no electrons are ejected. Work was needed. Einstein's hands drifted up towards his full head of hair.

When Einstein got stuck on a problem, he would play the violin (his 'Lina', as he affectionately called her). After countless performances of every score he possessed, he had no choice but to conclude that light had to consist of particles. He called these particles 'light quanta' – it was only later that they were popularly renamed 'photons'. This was music to his ears! Because by assuming that light essentially consisted of particles, he could instantly account for the photoelectric effect. Let us explain.

An electron is ejected when it absorbs a photon (yes, just one!) that carries enough energy to dislodge it. If the energy of the photon is too low, the electron simply stays stubbornly in place. That single photon – said Planck – always has an energy equal to $h \cdot v$. The higher the intensity of the light, the more photons, and therefore the more electrons are released. So Einstein's hypothesis immediately explains two key observations. First, the energy of the ejected electrons is independent of the intensity. Second, there is a minimum frequency required to eject an electron – namely the frequency corresponding to the binding energy of the electron. In other words: the energy doesn't gradually build up in the electron until it has enough to break away from the material. No. The energy of a photon is always absorbed in full by a single electron, or not at all. If a single photon doesn't carry enough energy, the electron doesn't budge.

But Einstein's vision found no favour. No one believed him, least of all Robert Millikan (1868–1953). Millikan spent over a decade trying to experimentally disprove Einstein's hypothesis, only to concede, grudgingly, that Einstein was right. For his own experiment – oh, the irony – he was awarded the Nobel Prize two years after Einstein.

FROM THE COMFORT OF YOUR ARMCHAIR: THE PHOTOELECTRIC EFFECT

The photoelectric effect is not just a pinnacle of scholarly achievement. It lies at the root of an astonishing number of technological gadgets that we rely on every day. A TV remote control, for example. The control sends infrared light towards a sensor in the television, electrons are released, then measured – and zap! Lights that come on automatically when it gets dark? The photoelectric effect! Automatic doors? The photoelectric effect! Heart rate monitors in your watch? The photoelectric effect! A broadband internet connection? The photoelectric effect!

The wave nature of light had been established for some time. Following the interference experiments of Thomas Young[1] (1773–1829) in 1803 and James Clerk Maxwell's theoretical description of electromagnetic waves in 1873, no one remained in any doubt about it. But by applying Planck's 'trick' in a very specific way, light began to reveal its particle-like side. It wasn't either/or, it was both/and.

But that wasn't the end of the affair for Einstein. It wasn't just light that consisted of particles. Everything that vibrates, he concluded in 1907, consists of particles. With this key insight, he went on to unlock a host of other experimental mysteries. Einstein capitalized on his discovery by means of a question very similar to Planck's black-box mystery: to what extent does the heat

---

[1] Thomas Young is sometimes described as 'the last man who knew everything'. His experiments with light showed that light consists of waves. In another capacity, he was instrumental in deciphering the Rosetta Stone.

capacity of a material (the amount of energy that is needed to heat it by one degree Kelvin) depend on its temperature? When a material is heated, the atoms in the material begin to vibrate. By assuming that these vibrations are both a wave and a particle (in this case phonons, from the Greek *phone*, meaning sound), he was also able to explain the mysterious experimental results surrounding them. Once again, it all came down to quanta. In fact, this insight is even more impressive than the discovery that light consists of particles, because phonons are an *emergent* phenomenon (we'll come back to this in Chapter 8), making them much more abstract than photons – and much less easy to define. The verdict was clear: the laws of classical physics were in desperate need of a rewrite.

Besides being an irrepressible genius, Einstein could also be a total pain in the neck. Over time, he made it his mission to play devil's advocate by constantly bombarding scientists with paradoxes and mind-bending questions. But due to – or rather, thanks to – his tireless attacks on quantum theories, he forced others to continually develop new arguments to counter his criticism and scepticism. He kept everyone thinking at the cutting edge, pushing them to rethink and reimagine not only their ideas but the very foundations of science itself. In this regard it is striking to note that Einstein, like Planck, initially struggled to accept the implications of his own discoveries. Planck, for his part, took his sweet time recognizing the value of Einstein's work. When proposing Einstein for membership of the Prussian Academy of Sciences in 1913, Planck dryly remarked: 'That [Einstein] may sometimes have missed the target in his speculations, as, for example, in his hypothesis of light quanta, cannot really be held too much against him.' Their contemporaries too were initially at a loss about what to do with these new quantum quirks. While physicists sensed that the ideas of Planck and Einstein marked the dawn of a new era in physics, few were prepared to abandon the solid ground of classical physics, which had been the gold standard for so long. That is the story of science: when you discover something new, you can seldom/rarely/never know in advance what new insights that discovery will lead to. But reality, stubborn as it is, will eventually catch up with you – whether you like it or not. In the end: quantum may have been the brainchild of a few eccentric geniuses, but their outlandish ideas did lead to a revolution.

IS $h \cdot v$ SEXIER THAN $E = mc^2$?

Most of us know Einstein mainly thanks to the third and fourth of his famous 1905 articles, which introduced the special theory of relativity and the iconic equation $E = mc^2$. Essentially, in these articles Einstein dismantled Newton's ideas about absolute time and space. According to Einstein, the speed of light is always constant, regardless of the relative speed of the observer or the light source. Picture this: if you're cycling alongside a beam of light, pedalling furiously, and you measure its speed, will it be slower than if you measure it while calmly standing beside your bike, sandwich in hand? Or, let's go further: if you shift into high gear and push the limits, might you eventually catch up to that beam of light? The answer, in both cases, is no. Light always travels at the same speed, and nothing – not you, not anyone, not anything – can ever outrun it.

In due course, Einstein's special theory of relativity would accord him superstar status. But was it truly the most revolutionary discovery since Newton? Not according to Einstein himself. He regarded his first paper, on the photoelectric effect, as more ingenious than his work on relativity. Why? Because he believed that if he hadn't developed the theory of relativity, someone else was guaranteed to have done so. Of course, black holes and time travel appeal much more to the imagination of the average mortal, but Einstein's explanation of the photoelectric effect was far more advanced and ultimately led to many more breakthroughs. So maybe it's time to give the T-shirts and coffee mugs a makeover: swap out the ubiquitous $E = mc^2$ for the truly monumental $E = h \cdot v$.

Einstein: 'Life is like riding a bicycle. To keep your balance, you must keep moving.'

84   *Why Nobody Understands Quantum Physics*

3.3 The first atomic models

Niels Bohr

Someone else who, like Planck, became close friends with Albert Einstein over time was Niels Bohr (1885–1962). In 1911, Bohr was fortunate enough to find himself at Cambridge as a young doctoral student in physics. True to form, he was on the lookout for interesting problems to sink his teeth into. At Cambridge he joined the research group of Joseph John Thomson (1856–1940), whose model of the atom (often described as a plum pudding) helped lead to the discovery of the electron. Thomson pictured the atom as a positively charged syrupy mass filled with negative electrons (the 'currants'). The sum of that sticky positive mass and its criss-crossing negative currants was a neutral atom. J. J. Thomson was awarded the Nobel Prize in 1906 for his discovery that electrons are particles. Over twenty years later, his son George received the Nobel Prize for his experiment showing that electrons are waves. Internal rumblings in the firm of Thomson & Son . . .

Bohr didn't mean to be a nuisance, but he still had reservations about Thomson's theory. He thought the whole 'pudding' analogy was far too simplistic. The proof of the pudding is in the eating. Long story short: theory and practice did not agree in the slightest. So there you had it: an interesting problem. Thomson, however, was less enthused. If Bohr didn't like his pudding, he'd better find himself some other colleagues. So in 1911 Bohr upped sticks and left for Manchester. A blessing in disguise, as it turned out. There he met his scientific soulmate Ernest Rutherford (1871–1937), an empiricist to the core, who would experimentally dismantle Thomson's pudding model down to its last currant. Under Rutherford, the structure of the atom was given the cachet it deserved: he compared the atom to no less a model than the solar system.

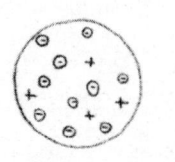

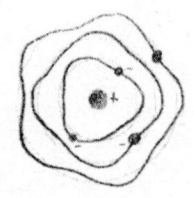

The atomic models of Thomson (left) and Bohr (right)

By putting Thomson's model to the experimental test, Rutherford eventually arrived at his own model of the atom. The pivotal experiment involved firing alpha rays (helium atoms stripped of their electrons) at a thin sheet of gold foil to see how the rays would be deflected. What did he see? Almost all of the alpha rays passed easily though the foil. Some flew straight through, others were slightly 'scattered' (or deflected), while a very small minority (0.000001 per cent) bounced straight back. Rutherford didn't know what he was seeing. But whatever it was, he could only compare it to a cannon ball being reflected back by a paper handkerchief.

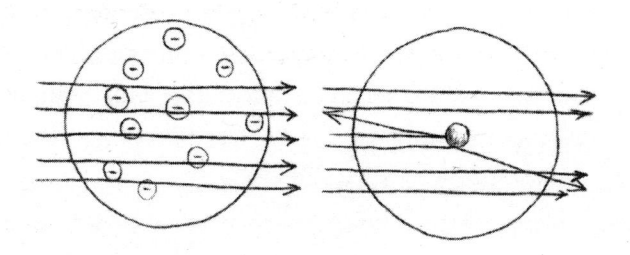

Thomson's scattering versus Rutherford's

After weeks of research, the answer began to dawn on Rutherford. If an atom really were a pudding full of currants, the probability of an alpha particle being reflected straight back would be zero point zero. The electron currants were far too light, and the pudding far too syrupy, to put up any resistance. The overwhelmingly powerful alpha particles should have zipped through without exception. The only explanation Rutherford could come up with, after extensive thinking and tinkering, was that an atom had to be virtually empty, with nearly all its mass concentrated in a tiny, dense, positively charged core. That minuscule piece had to be the atomic nucleus.

Rutherford's experiment revealed something astonishing: the diameter of a gold atom nucleus had to be roughly 10,000 times smaller than the diameter of the whole atom. If an alpha particle happened to strike that tiny positively charged nucleus (a probability of just 0.000001 per cent), it would indeed bounce back. But hold on, that would mean the space is 99.9999999999 per cent empty? And if matter is primarily empty space, how come it feels solid? Why don't we sink through our chairs? Why don't our boxing gloves pass through the punch bag? Rutherford's atomic model turned conventional wisdom upside down. This New Zealander had such exceptionally powerful antennae (like Einstein) that he simply *knew* there was more to uncover. Surely the story of the atomic nucleus couldn't end there. The atomic model that Rutherford introduced, comparing the positive atomic nucleus to the sun, with negative electrons revolving around it in fixed orbits like planets, became a cornerstone of twentieth-century physics.

Of course, every solution brings new problems. How to explain the fact that electrons circling the nucleus do not lose their energy? Accelerated electrons constantly emit electromagnetic radiation, so shouldn't they eventually run out of energy and crash into the nucleus? Apparently not. But why not? How come the (negative) electrons don't stick to the (positive) nucleus like tiny magnets? This perplexing paradox made one thing crystal clear: classical physics would not offer the solution. Planck's constant could no longer be ignored.

Niels Bohr, a theorist through and through (unlike Rutherford, a born-and-bred experimentalist), was deeply impressed by this experimental firepower. He took a closer look at Rutherford's experiments and translated his colleague's atomic model into a rigorous mathematical theory. Bohr blended a bit of Planck, a dash of Einstein and a splash of audacious assumptions, gave it all a vigorous shake in the classical physics mixer, and arrived at a hypothesis that could finally explain nearly the entire workings and structure of the atom.

Whereas Einstein and Planck had discovered that the energy of a light particle is quantized, Bohr went a step further by assuming that electron orbitals must also be quantized. Electrons are not like currants scattered in a pudding; instead, they travel along well-defined paths ('orbitals') around the atomic nucleus. According to Bohr, each of these orbitals has a specific energy level. The electron orbital closest to the nucleus is like the fundamental tone of our guitar string: it has the fewest vibrations (or 'nodes') and thus the lowest energy. Orbitals further from the nucleus correspond to higher multiples of this fundamental energy level. The further from the nucleus, the more energy the orbitals have. Just like the harmonic tones of a guitar string, these energy levels fit neatly together, creating a harmonious quantum system.

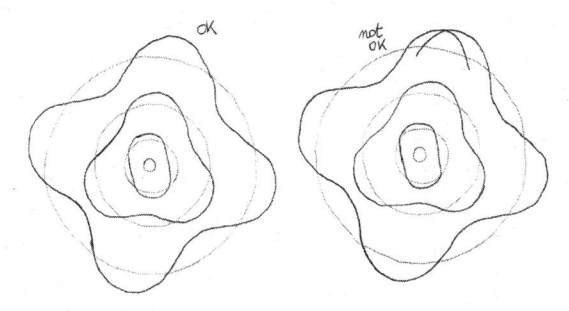

According to Bohr's atomic model, an electron follows a path that loops around the nucleus like an upright wave with a very precise number of undulations. Waves that do not complete a perfect full rotation (right-hand illustration) do not occur.

When an electron, calmly cruising along its orbital path, encounters an incoming light particle (photon), it gets all excited (which is actually the technical term). It eagerly absorbs the light particle, gaining the energy boost it needs to jump to an orbital with a higher energy level. But this only happens if the light particle (with energy $h \cdot \nu$) contains exactly the right amount of energy – equal to the difference between the energy of the orbital of departure and that of its destination. After all: in nature, nothing is ever lost. When the electron later drops back to a lower-energy orbital (a process that happens entirely at random), the surplus energy is released as a photon, emitted with that exact energy. As long as the electron keeps dutifully spinning in its orbital, it doesn't emit any energy (or photons). Much like a ladder, an electron can never be found between two orbitals. It can only 'stand' on the defined rungs. No 'half-rungs' or 'in-between rungs' are allowed. A ladder or a flight of steps, like the energy levels in an atom, is quantized. Each atom, in fact, has its own structure of energy steps.

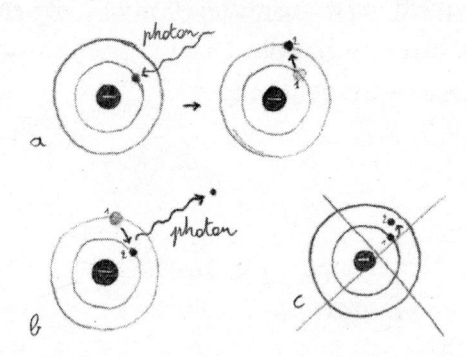

(a) When a photon is absorbed, an electron is propelled to a higher energy orbital; (b) if the electron drops back to a lower orbital, it releases a photon; (c) the electron can never end up somewhere in between two orbitals.

But what, then, is this mysterious force that keeps electrons spinning in their orbitals around an atomic nucleus instead of spiralling into it? On this question, Bohr had no answer. The world would have to wait until 1925, when Schrödinger and Heisenberg finally shed light on the matter. Still, Bohr was the first person to present a very clear atomic model, which instantly gained him worldwide renown. This model not only explained why atoms emit light at specific frequencies, but also laid the groundwork for understanding the periodic structure of Mendeleev's table.

The major 'issue' with Bohr and Rutherford's theory is that they still treated the electron as a particle that, for some mysterious reason, travels in fixed orbits around the atomic nucleus, like the planets around the sun – and is capable of jumping between orbits at certain frequencies. They never even imagined that an electron could be both a particle and a wave, and that this wave gives shape to the atom. Conceptually, the wave nature of an electron is far harder to grasp than that of light. Like

Planck before him, Bohr presented his theory as a convenient 'trick', a pragmatic solution to make the numbers work. But does that really matter? What counted most was that it worked – the end justifies the means. However, like Planck, Bohr could not yet provide a satisfying answer to the deeper 'why' behind it all.

BEER AND THE BOHR BOYS

Q: What is the link between Niels Bohr and (probably) the best beer in the world?
A: The Royal Danish Academy of Sciences and Letters, an heir to Carlsberg.

Here's the story. The Academy was so honoured by Bohr's return to Denmark that it promptly handed him a bag full of money to establish his own institute of physics in Copenhagen, where he could hire the world's brightest students. To this day, Carlsberg taps off some of its profits directly to the Academy, fuelling scientific research and nurturing new talent.

At the centre of the brewery site stands the founder's villa, which should house the most prominent Danish scientist of the time. Bohr was among the first to receive this honour. His institute became a hotbed of new ideas, attracting the scientific *crème de la crème* – including Heisenberg, Pauli, Dirac, Gamow, Landau, Ehrenfest and more. These 'Bohr boys', ever thirsting for knowledge, laid the groundwork for quantum physics. Their vision is still referred to today as the 'Copenhagen School'. And Bohr? He was the great mentor, the godfather of quantum mechanics. So, if you want to support quantum physics, you know what to drink. *Skål!*

3.4 On particles and wave packets

In the pre-war, still-pristine France of the early twentieth century, chairs in experimental physics were as plentiful as grapes in the vineyard. But the First World War cut down more than just young men – it also extinguished careers, dreams and the vitality of physics itself. No matter their brilliance, everyone was sent to the trenches. In post-war France, marked by exhaustion and loss, there was little appetite to revive the sciences. In addition, an embargo had been imposed that barred French scientists from maintaining contact with their German and Austrian counterparts. All exclusive (or explosive) materials, whether bombs or intellectual property, were anxiously retained within national borders.

Prince Louis Victor Pierre Raymond de Broglie

Someone who had been spared the trenches was the French prince Louis Victor Pierre Raymond de Broglie (1892–1987).[1] His expertise

---

[1] The name 'de Broglie' causes a lot of confusion in terms of pronunciation. Some people (us included) prefer to pronounce it BRO-cli, while others swear by BRO-gli, although those in the know say it sounds more like a drawn-out BRO-O-OY.

had been put to use in the radio communications service (in the Eiffel Tower, of all places), where he was tasked with intercepting enemy radio messages. Yet de Broglie had not initially been destined for a life in science. He had studied medieval history at the Sorbonne and dreamed of becoming an ambassador. His father was resolutely opposed: 'Tut-tut, you will become a statesman, like me. Or prime minister, *pourquoi pas!*' History had other plans. Influenced by his brother, a natural scientist, young Louis traded Gothic cathedrals for cathode rays. He broke off his engagement, shelved his history books, severed his social ties and cleared out his phenomenal memory stuffed with the classics of French literature to make space for maths and physics. He would devote himself to science. And like the enduring Notre Dame, de Broglie's contribution to quantum physics would stand firm for centuries.

NOBLESSE OBLIGE

One fine day, a gentleman from across the Channel knocked on the door of Louis Victor Pierre Raymond, better known as the prince de Broglie, at his stately residence in Neuilly-sur-Seine, just outside Paris. The visitor, none other than the Russian-American George Gamow, was obliged to dust off his rusty French and master its tongue-twisters in an effort to understand his host and to make himself understood. The prince, you see, didn't speak English. And Gamow? Well, he *ne parlait* very good *le français*.

Fast forward a year, and the tables turned. The same French prince crossed the Channel to deliver a lecture on his scientific work. A marvellous explanation, presented in near-flawless

> English. The prince, after all, was a man of principle. He was more than happy to adapt. But back home, in France, one speaks French. *Voilà*.

It all began with de Broglie's 1924 thesis, '*Recherches sur la Théorie des Quanta*' (Research on the Theory of the Quanta). The examiners didn't know what to do with it. Not because they didn't understand it, but despite the fact they didn't understand it. All particles have a wave nature, the thesis boldly claimed. To most scientists of the day, the notion that particles could also be waves was as outlandish as serving Camembert baked in the box: totally off the mark. Things consist of particles, and light is waves (but occasionally a bit particle-like) – that was the general tenor at the time. Still, to play it safe, de Broglie's supervisor, Paul Langevin, quietly passed the thesis along to Einstein. Very informally, of course, because Einstein was and remained a German. 'Curious to hear your thoughts.' Cue Einstein, hands buried in his famously unruly hair for several months. When he resurfaced, he declared the work to be good. *Vachement bien*, in fact.

De Broglie had built on Bohr's work, but given it a radical new twist: if waves are particles, then particles must also be waves. *Bref*, all particles are waves. Electrons, protons, golf balls; everything is made up of waves. Or, more precisely, they're wave packets: localized bundles of waves formed by the superposition of waves with different frequencies. The extent to which the wave packet is localized depends on the wavelength with the greatest amplitude (its parent wave): it cannot be *more* localized than the latter's wavelength. According to de Broglie, every particle is a wave packet, smeared out across the entire span of the packet, existing simultaneously everywhere between its start (a) and end (b).

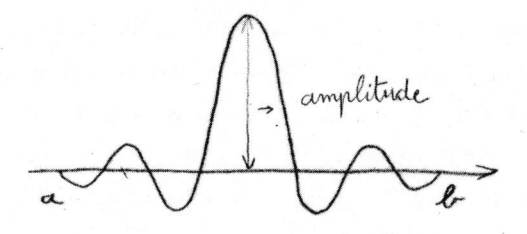

A wave packet. The arrow indicates the amplitude. The length of the wave packet depends on the parent wave (the wavelength with the greatest amplitude).

De Broglie's main contribution to quantum physics is a formula that expresses how to determine the wavelength of any particle, and thus how strongly it can be localized. To this end, he introduced the 'de Broglie wavelength', which is equal to Planck's constant divided by the momentum (mass times velocity) of the particle:

$$\lambda = \frac{h}{p} = \frac{h}{m.v} \simeq \frac{h}{\sqrt{m.k.T}}$$

The symbol λ refers to the length of the wave packet. For a system in equilibrium, the momentum is equal to the square root of its mass times Boltzmann's constant times the temperature.

This formula holds a prominent place in physics, as it allows us to determine when quantum mechanics becomes necessary to describe a system – whether we're tracking the trajectory of a golf ball, an electron in a metal or the hydrogen atoms in the sun.

In essence: when the average distance between particles is smaller than their respective wavelengths, their wave packets overlap. At this point, they lose their identity, they interfere, and then you've guessed it: this is quantum territory. If not, Newton's classical physics suffices, and quantum mechanics can stay on the bench. There are three situations in which the de Broglie wavelength exceeds the distance between particles: where there are many particles close together (the more particles, the closer they are packed); when the particles are very light (lighter particles have larger de Broglie wavelengths); and at very low temperatures (as temperature drops, particles lose energy, extending their wavelengths).

Particle	Temperature	Density	Mass	Q Factor
Golf ball	273 K	$10000/m^3$	$2.10^{24}u^*$	10^{-28}
Neon (gas)	273 K	$10^{25}/m^3$	$20u$	10^{-7}
Helium (gas)	273 K	$10^{25}/m^3$	$4u$	10^{-6}
Helium (liquid)	4 K	$10^{28}/m^3$	$4u$	1
Rubidium (BEC)	10^{-7} K	10^{19}	$87u$	1.5
Electron in aluminium	273 K	$10^{29}/m^3$	$5.10^{-4}u$	10^4
Neutron in neutron star	10^8 K	$10^{44}/m^3$	$1u$	10^6

*u is the unified atom mass unit, roughly equal to the mass of a proton.

Q is the size of the de Broglie wavelength divided by the average distance between particles. If Q is greater than 1, it's quantum. The greater the number, the more quantum it is. Electrons are always quantum. Golf balls and noble gases clearly are not.

In the world that we experience, it might seem as if quantum physics is irrelevant. But make no mistake, quantum physics is woven into the very fabric of our daily lives. After all, what makes matter solid? Quantum! What enables chemical reactions and bonds? Quantum! What gives the world its vibrant colours? Quantum!

3.5 The double-slit experiment

Would *Monseigneur le Prince* de Broglie be so kind as to demonstrate his bold claims from that mysterious yet ingenious doctoral thesis of his? After all, we'd love to see this wave-like behaviour of electrons in action. De Broglie knew there was no better way to demonstrate the crazy nature of quantum physics than with the 'double-slit experiment'. Except he had never conducted that experiment himself. That honour fell to Clinton Davisson (1881–1958) and Lester Germer (1896–1971), who performed the famous test in 1927 at Western Electric's labs (later known as the illustrious Bell Labs).

In their experiment, electrons were fired one by one at a crystal lattice of nickel, with the crystal playing the role of the two slits. Behind the crystal stood a projection screen, ready to capture where the particles landed. According to de Broglie, an interference pattern should appear, because when two waves intersect, they interfere. This interference can either amplify (waves adding together) or diminish (when waves partially or fully cancel each other out) the resulting amplitude. Time to put theory to the test.

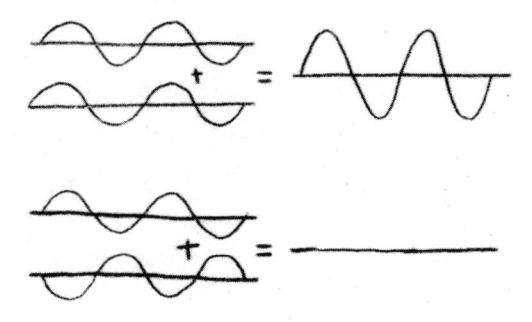

Positive and negative interference

To showcase the importance of the double-slit experiment, let's first try it with something undeniably particle-like (say, tomatoes), then with waves that are unambiguously wave-like (water waves) and finally with electrons.

First the tomatoes (a). Suppose we cover the left-hand slit. Well-aimed tomatoes fly through the right slit, leaving a long red smear of splattered tomato guts on the right side of the projection screen. If we cover the right slit instead, the tomatoes fly through the slit on the left, creating a similarly messy trail on the left side of the screen. But if we leave both slits open, two overlapping tomatoey trails appear on the screen. The final result is simply the sum of the two individual outcomes. Conclusion: tomatoes are particles. No doubt about it.

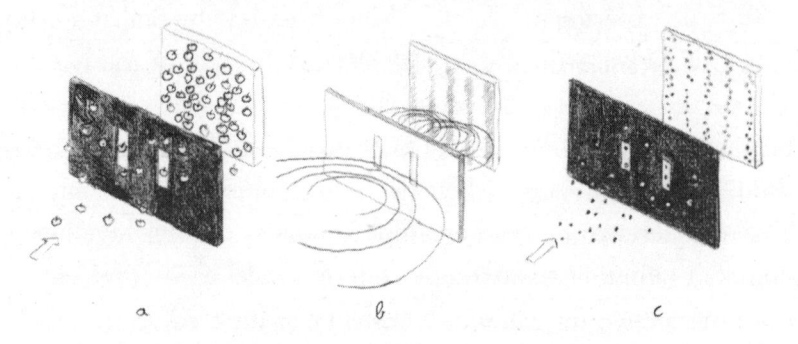

The double-slit experiment conducted with (a) tomatoes, (b) water waves and (c) quantum particles (electrons).

Now let's conduct the experiment with water waves (b). Drop a pebble into the water and ripples spread outward across the surface. We repeat the scenario: cover the left slit, cover the right slit and finally leave both slits uncovered. This time, things get a bit more interesting. When the left slit is covered, the wave passing through the right slit bends and spreads out (a phenomenon called diffraction), creating a series of smaller ripples on the other side of the screen. If the right slit is covered, the same happens through the left slit: the wave bends, diffracts and creates a pattern of smaller ripples beyond the barrier.

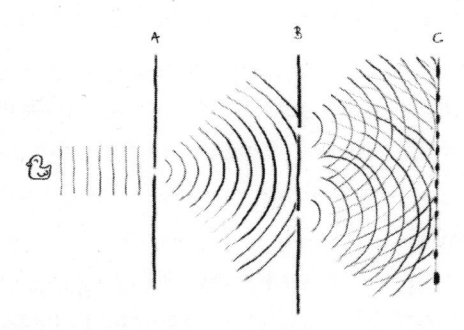

Interference occurs when the wave, after passing through the slit in the first wall (A), 'breaks' on passing through the double slits (B), and the broken waves meet again to the right of wall B. Either the waves strengthen each other, or they weaken each other. This effect creates interference patterns on the projection screen (C).

When waves ripple through both slits simultaneously, the resulting 'broken' waves overlap.

This superposition of waves leads to interference: the waves either amplify each other (constructive interference) or cancel each other out (destructive interference). The final pattern on the projection screen depends entirely on how the two wave streams, originating from each slit, interact. Conclusion: waves behave as waves.

Now for the grand finale: electrons (c). We fire them one by one (a crucial detail!) at the screen. Given that electrons are (quantum) particles, you would expect the same outcome as in the tomato test. But surprise: electrons produce the same end result as the test with the water waves. De Broglie was right. If an electron does indeed pass through both slits simultaneously, it is spread out over its entire wavelength. The crucial factor here is the distance between the slits. If it is smaller than the de Broglie wavelength of the electron, the electrons behave like waves; if the distance between the two slits is larger (as with the tomatoes), they behave like particles. The difference between a particle and a wave is, in short, its de Broglie wavelength.

And now, for the cherry on top: a slight tweak with significant consequences. If we position a detector at one of the two slits so we can see which of the two the electron flies through, the interference pattern disappears. The electron reverts to behaving like a particle. It couldn't be any clearer or more mysterious. Observing something changes its character. Richard Feynman (a key figure in Chapter 6) aptly described the double-slit experiment as a phenomenon 'which is impossible, absolutely impossible, to explain in any classical way, and which has in it the heart of quantum mechanics'. Everything in quantum can be traced back to this experiment. Grasp this, and you're on your way to understanding (almost) everything.

3.6 Heisenberg's microscope

> *Take up your microscope, you fool, and tremble.*
> *The same abyss with the same waves is all around.*
> *In the infinitely small the same worlds abound*
> *As in the infinitely large. [. . .]*
> *The limitation lies not in nature, but in*
> *The crude instrument, the imperfect organ;*
> *That eye of yours is less a means than an obstacle;*
> *You must magnify the gaze to magnify the spectacle;*
> *Within the small is the immense.*
>
> — VICTOR HUGO

The double-slit experiment leads us seamlessly to Werner Heisenberg (1901–1976) and his microscope. Here, the 'microscope' symbolizes not just an instrument but the very act of observation itself. What does that actually mean, to observe? What does it mean for something to have a property? When we think we understand something, what do we actually know? If we want to study something, we first need to look at it. But in order to see something, we need light. And when we shine light (photons) onto a system, that very act inevitably disturbs the object we're trying to observe.

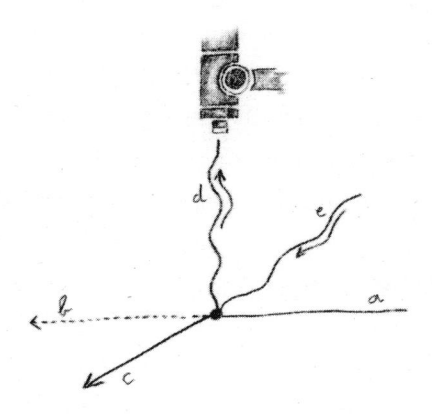

The Heisenberg microscope: a, b: original trajectory of the electron; c: scattered electron; d: reflected light particle; e: incident light particle.

This idea calls for a bit of nuance. Simply looking at a macroscopic object, like a clock, doesn't alter its state. Checking the time won't make the clock hands jump around. But when it comes to the quantum realm, things are different, particularly if we're trying to pinpoint the exact position of an electron. The disturbance caused by observation depends on how the momentum of the light particles compares to the momentum of the object being observed. Think of it like a bumper car: if one car collides head-on with another stationary one, momentum is transferred, sending the second car flying. Contrast this with a fly bumping into a windshield: the effect is negligible. Size and momentum are everything. Suppose you want to measure something with extreme precision, say, down to 10^{-10} metres. For that, you'll need light with a matching wavelength of 10^{-10} metres. But here's the catch: the momentum of the light particles (photons) at this wavelength is comparable to the momentum of an electron. This means shining such light on an electron will disturb it significantly, akin to a bumper car collision. A golf ball, by comparison, has a momentum that is much greater, and will therefore be unaffected, just like the windshield in our earlier analogy.

This line of thought led Werner Heisenberg to formulate his famous uncertainty principle: the more precisely you determine the location of an object, the more you disturb it.

This principle also provides an intuitive explanation of de Broglie's wavelength of a matter particle: it corresponds to the wavelength of a light particle whose momentum is equal to the momentum of the matter particle. In other words, if you observe a matter particle using light with a wavelength equal to its de Broglie wavelength, you will significantly disturb that particle. A golf ball has a de Broglie wavelength of 10^{-20} metres. To have any measurable effect on it, you would need light with a wavelength of at least 10^{-20} metres. However, light with such an astronomically small wavelength has so much energy that it will instantly scorch not only your golf ball, but virtually everything in its vicinity. On the other hand, an electron has a much larger de Broglie wavelength than a golf ball, meaning that even ordinary light will inevitably disrupt it. In the quantum world, to observe is to disturb.

Heisenberg features prominently again in the next chapter. By then, inspired by his own thought experiment with the microscope, he will formally derive his famous uncertainty principle. As a teaser: the more precisely you know the position of an object, the less accurately you can know its momentum (mass times velocity), and vice versa. In other words, it's impossible to know the position and momentum of a microscopically small object at the same time. This idea runs completely counter to the laws of classical physics and/but/so is the absolute essence of quantum physics. It is the only consistent way to reconcile the quantum theory of light with our understanding of electrons. With his thought experiment, Heisenberg had already shifted the conceptual landscape, but with his uncertainty principle, he established one of the weightiest cornerstones of quantum physics.

Werner Heisenberg

IN A NUTSHELL

> Quantum particles are described by waves and their superpositions. Give me your wave function, and Schrödinger will evolve it.

> A wave function is not really a wave. It does not describe reality. It is information. It represents the probability that a measurement will produce a certain result. A whole lot in quantum physics depends on randomness.

> Superposition. In quantum land, things are not either/or, but both/and.

> *Dramatis personae*: Erwin Schrödinger, Max Born, Werner Heisenberg, Wolfgang Pauli, Paul Dirac.

FOUR

THE FIRST QUANTUM REVOLUTION

4.1 Waves of whimsy

> *I love it when a plan comes together.*
> – HANNIBAL, *THE A-TEAM*

Something was brewing at the start of the twentieth century. A wave of assumptions, suspicions, hypotheses and 'what ifs' swept over the scientific community. What *had* become clear was that the microscopic world was not simply a reflection of the macroscopic one. Numerous theories suddenly gained an air of mystery, with particles that could simultaneously be everywhere and nowhere, particles that were waves and waves that ceased to be waves once they were measured. How was anyone supposed to make sense of it all? But the world kept spinning, the war came to an end and one day the year 1925 arrived. From that moment on, discoveries cascaded like falling dominoes, opening up new, unexplored ways of thinking. Experiments that still seemed impossible in the previous chapter could now be carried out and explained; hypotheses and suspicions gave way to probabilities and likelihoods. Gradually, abstract physics began to reveal its true nature. Physics was dead. Long live quantum physics!

Erwin Schrödinger

We begin this chapter with Erwin Schrödinger (1887–1961), although we could just as easily have started with Werner Heisenberg or Max Born or Paul Dirac. Around 1925, there was hardly anyone who *wasn't* working on the recently discovered concept of wave–particle duality. It's therefore no coincidence that within a short space of time, and almost simultaneously, various explanations for this mystery emerged. Yet, chronology matters less than understanding how these theories came to be, and why they're so crucial.

The revolution started on several fronts at once. Schrödinger picked up where de Broglie left off: every particle exhibits wave-like behaviour. That's a very noble basis from which to start, but the question remains: how can this wave, and its evolution over time, be described in mathematical terms? Put simply: what is the wave equation?

In Chapter 1, we saw that waves (composed of many classical particles) can be described using matrices and differential equations.

Schrödinger took the bold step of applying these same equations to describe the properties of a single quantum particle. He introduced a wave function, denoted by the Greek letter psi (ψ), which he associated with that single particle. This wave function is analogous to the wave function of a vibrating string, which describes its distance (amplitude) to the x-axis at any point on the x-axis and at any moment in time (t):

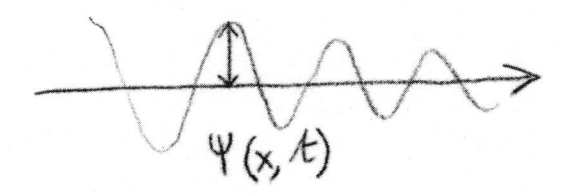

$$\psi(x, t)$$

In order to be consistent with Bohr and de Broglie's theories, Schrödinger had to make a number of crucial modifications to the classical wave equations. A first major change is that Schrödinger uses complex numbers instead of real numbers for the quantum wave function. Assuming that the whole space is filled with points, the wave function assigns a complex value to each of these points.

Additionally, Schrödinger let the wave equation evolve over time in a different way (in case you were wondering: according to a first derivative to time instead of a second one). After doing the necessary calculations and fine-tuning, he eventually penned the most important formula of quantum physics: the Schrödinger equation.

Despite our intention not to include any formulas in this book, we will make another exception for the Schrödinger equation. Not only because it is so revolutionary, but also (and this is probably a bit more surprising) because of its mathematical aesthetics. For Einstein is right: a theory that is correct must also be beautiful, in the sense that it enlightens and inspires by revealing a profound truth about nature. Thus, the Schrödinger equation:

$$i\hbar \frac{\partial}{\partial t}\Psi(x,t) = -\frac{\hbar^2}{2m}\frac{\partial^2}{\partial x^2}\Psi(x,t) + V(x)\Psi(x,t)$$

i = square root of -1 (imaginary number)
$\hbar$ = Planck's constant (h) divided by 2π
$\Psi(x,t)$ = wave function with a complex value at each position (x) and time (t)
$\frac{\partial}{\partial t}$ = derivative of the wave function to time (how it changes as a function of time)
m = mass of a particle
$\frac{\partial^2}{\partial x^2}$ = second derivative of the wave function to its position
$V(x)$ = external potential (potential energy of the system at position x)

The Schrödinger equation is essentially a matrix equation in infinite dimensions (the Hamiltonian) that sets out how the complex numbers of the wave function evolve over time at each point in space. With the Schrödinger equation, everything changed. It allows us to calculate electron orbitals and their corresponding energy. For the aficionados: the orbitals are the eigenvectors (stationary states) of the matrix we call the Hamiltonian, and the energies of those orbitals are the eigenfrequencies.

However, the Schrödinger equation has more tricks up its sleeve. It can describe how non-stationary states evolve (when an electron is in a superposition of different orbitals that evolve over time), how light interacts with electrons and how electrons interact with each other. In short, the Schrödinger equation marked the beginning of a brand-new era. In this sense, everything that follows in this book, at least from a mathematical point of view, is either directly or indirectly an application of this one formula. What Schrödinger couldn't have foreseen is that others, not even much later, would use his magic formula to describe entire materials. And superconductivity. And even nuclear forces. All matter, in fact. As one of Schrödinger's many lovers teasingly remarked: 'When you began this work you had no idea that anything so clever would come out of it, had you?'[1]

And yet, a grey area lingered. Wave functions could be distilled from all those imaginary and complex numbers, and all the information about a particle could be deduced from the wave function, but the real question remained: how exactly should we interpret this information? Finding formulas that describe *how* everything works isn't quite the same as explaining *why* something works. And to be honest, Schrödinger still wasn't entirely sure what exactly a wave even was. The answer was pulled from the hat of another member of the quantum family: Max Born.

DISHING THE DIRT (1/2)

Schrödinger garnered fame for his experiment with the cat (Chapter 5), but his private life had a very dark side: there are credible allegations that he had relationships with underage girls. When asked about his state of mind, he would answer: chronically in love. His unconventional love and lifestyle earned

---

[1] Erwin Schrödinger, *Collected Papers on Wave Mechanics*, Minkowski Institute Press, 2020, p. xxi.

him plenty of criticism, not to mention a whole brood of illegitimate children. Nevertheless, his sense of style and etiquette were impeccable, his knowledge of psychology, Eastern philosophy and literature untouchable, his interest in fellow human beings, well . . . inexhaustible, and his English verged on perfection (he was also fluent in French, Spanish, Italian, Greek and Latin). Oh, and he was also an accomplished poet. It's an open secret that his wave equation was conceived during a two-week retreat somewhere in the Swiss Alps. The fact that he was in the company of a woman comes as no surprise – though her identity remains unknown. In the months following his famous entanglement, he wrote no fewer than six other papers that ultimately formed the true basis of quantum physics. It's all a matter of inspiring, and being inspired . . . With his 1944 book *What is Life?*, in which he endeavoured to unravel the great mysteries of biology on the basis of physics, Schrödinger was incidentally responsible for the fact that many budding physics students – including James Watson and Francis Crick, the eventual discoverers of DNA – changed their study paths and went on to study (molecular) biology.

Just like Planck and Einstein, Schrödinger grappled with the philosophical implications of his own findings – a struggle that was a constant in the early years of quantum physics. He preferred to stay true to the traditional methods of classical physics. But as it became increasingly difficult to ignore quantum theories, he faced a fait accompli: abandon old principles and rethink the bigger picture. In this way, Schrödinger found himself in the same camp as Einstein, whose main aim, if at all possible, was to demonstrate that quantum physics could not be the final word – its predictions being too crazy to be true.

4.2 Waves of information

> *A possible experience or a possible truth does not equate to real experience or real truth minus the value 'real'; but, at least in the opinion of its devotees, it has in it something rather divine, fiery and high-flung, a constructive will and conscious utopianism that does not shrink from reality but treats it, on the contrary, as a challenge and an invention.*
> — ROBERT MUSIL, *THE MAN WITHOUT QUALITIES*

Max Born (1882–1970) had read Schrödinger's papers and, like everyone who kept up with the times, was fascinated by the question: particle *or* wave? Particle *and* wave? It was Born who realized how the Schrödinger equation should really be interpreted. And – to be honest – he was also a bit frustrated. With himself. Because he could have been the one to discover it. Born was simply unlucky not to possess the same sharp intuition as Schrödinger. He was far too much of a maths whizz for that. Still, we owe it to Born that, when we make a measurement, we also know *what* we are measuring, and how to interpret it using a wave function. So, what is that 'what', then? It is a probability. That might sound odd, but in quantum physics, it's not all that peculiar. Quantum physics is merely an attempt to mathematically represent everything that happens at the micro level. After all, a wave function is nothing physical; it is not reality. It contains the information *we* have about that reality; it relates to information about the probability of a particle being here or here (or there) when measured.

THE UNBEARABLE LIGHTNESS OF UNCERTAINTY

Einstein swore by the deterministic theories of Descartes, Newton and Laplace: if something is this way or that at a certain moment in the present, something will be this way or that in the future. This led Einstein to his famous saying that God does not play dice, meaning that there's no such thing as randomness. In classical physics, randomness is merely a reflection of our ignorance – a consequence of our inability to measure the exact position and speed of every individual particle, given their sheer number.

In quantum physics, however, the situation is entirely different. Here, chance and probability lie at the heart of reality, forcing us to abandon the deterministic worldview. Measurement outcomes are fundamentally random: the wave function only provides the probability of obtaining a particular result. Einstein's discovery that something can be both a wave and a particle positioned him as one of the founding figures of quantum physics. Yet he struggled to reconcile the implications of the quantum theories that followed from the work of de Broglie, Schrödinger and Heisenberg, as they clashed profoundly with his worldview.

There is, of course, a sound technical side to Born's theory. For simplicity, let's assume that the wave function consists of real numbers. The value of the wave function can be less than zero (when it goes below the axis), equal to zero (where it crosses the axis), or it can be positive and therefore greater than zero (when it goes above the axis). You are most likely to find a particle where the wave function is greatest (or: the amplitude is greatest). However, there's a subtle issue: probabilities cannot be negative, but a wave can. Born didn't see that as problem. In fact, he found a solution: you can obtain a positive outcome by always taking the square of the wave function.[1] And that square represents the probability of finding the particle at a particular spot if you were to measure its position.

Born's probability interpretation brought clarity to the elusive concept of electron 'orbitals'. Electrons, like other quantum particles, are not located at a single, fixed position; rather, they exist as probabilities spread across space – everywhere and nowhere simultaneously. *Voilà*, there you have it. This idea is much easier to understand if, as stated earlier, we stop thinking of electron orbitals as orbits and instead think of them as a wisp of mist.

Thanks to the wave function, we can, very concretely, calculate the probability distribution of the outcome of every possible experiment. The wave function encapsulates the knowledge we, as observers, possess of the quantum system in question. On the other hand, each new measurement also yields new information, meaning that we have to update our knowledge with each new measurement. This, in turn, causes the collapse of the original wave function. This collapse is unavoidable and also irreversible. The more data you gather, the fewer superpositions of possibilities remain. What remains is a broken superposition, a curtailed wave function and therefore a very 'classical' back-to-square-one state.

---

[1] If the wave function consists of complex numbers, calculate the probability by taking the absolute value of the complex number squared.

Max Born

4.3 Double slits: the theory

Thanks to de Broglie, we know that particles are waves. Now, prophesizing that particles and waves reinforce or cancel each other is one thing. But providing an explanation for this in mathematical terms is quite another challenge altogether. Because what is interfering with what, exactly? And how do we calculate that? Schrödinger and Born's theory provides the answer to this question. In the double-slit experiment, an electron detected at position x on the projection screen may have taken one of two paths: through the left slit or the right slit. Each of those paths contributes to the wave function at this position ($\Psi_L(x)$ and $\Psi_R(x)$). According to Born, the probability of encountering an electron at position x is the square of the sum of these amplitudes. This can be written as: $|\Psi_L(x) + \Psi_R(x)|^2$ Remarkably, this probability can be zero, even if the two separate wave functions are not zero, provided they have opposite signs (plus and minus). This is interference. And this explains the interference pattern of the double-slit experiment.

But Schrödinger's equation, paired with Born's interpretation, reveals something even more startling: why the interference pattern disappears if you try to see which slit an electron has passed through. Because if you do, the outcome is just like the tomato experiment. *Que?* How can observation itself affect particles and cause the wave function to collapse like a house of cards?

The reason is that, when we shine a light source on electrons to observe them – bearing Heisenberg's microscope in mind – they entangle with the light particles (photons). This entanglement (or superposition) enables us to better observe the electron, but we also disrupt the interference. You can no longer add the amplitudes together. It's almost as if the photons are 'peeking' to see which slit the electron goes through, inadvertently conducting a measurement on the system. According to Born, this is precisely what triggers the collapse of the wave function. Think of it as an extension of the Pauli effect: if meddling particles get too involved, the experiment falls apart. In the double-slit experiment, this interference means the particles revert to behaving classically – like tomatoes. Another, complementary way of explaining the act (and impact) of a measurement is that the electron entangles with the photon used to observe it, preventing the wave functions from combining cleanly.

Wait. How did scientists ever observe interference in the first place? Easy: by *not* looking. That is to say, by not looking at the slits (the process), but only at the end result (the projection screen). Conclusion: the world is really not what you see. Reality is infinitely more fascinating.

The double-slit experiment continues to fascinate us to this day. Where exactly does the boundary of 'big' lie? How big does something have to be to stop behaving like quantum? And what's the deal with buckyballs? Do they behave like tomatoes or electrons? A buckyball is a molecule that consists of sixty atoms and, because it constantly absorbs and emits photons, interacts very strongly with its environment. Given these properties (big size, a lot of interaction), the claim that a buckyball's interference could, on a good day, also be demonstrated experimentally was firmly rejected.

Never say never, thought a couple of experimentalists (Markus Arndt and Anton Zeilinger, who were colleagues from Vienna), who cordially defied all the theorists who claimed that this was not possible. After all, experimentalists have always held a slight edge in the hierarchy of science, no matter what the *Homo theoreticus* might claim. That's Stevin's legacy.

The experiment took place in 1999. And indeed, it worked: the clearly visible interference patterns didn't lie. Faced with this, the theorists who had dismissed the idea as impossible scrambled to refine their arguments. They pointed out that the photons emitted by the buckyballs had wavelengths larger than the distance between the two slits. Of course! (As if they hadn't considered that before.) This meant the photons couldn't reveal which slit the buckyball passed through and therefore couldn't trigger a wave function collapse.

If we were to observe the buckyballs non-stop with light with a wavelength smaller than the distance between the slits, no interference would occur at all. The result is clear: the double-slit experiment provides the best evidence that Born's interpretation of the wave function is correct. Wave functions are receptacles of information and probabilities.

4.4 Quantum tunnelling

Another, equally fascinating consequence of the wave nature of quantum particles is quantum tunnelling, or the tunnel effect. This is certainly a strange beast. But one that solidified belief in quantum, since it's the only theory able to explain it. Where classical intuition tells us that particles with low kinetic energy cannot surmount a high barrier and should therefore be reflected, quantum particles can, in fact, 'tunnel' straight through – even when they lack the energy to overcome the barrier.

It's as if a cyclist, instead of climbing over a first-category col, somehow rides straight through it. Counterintuitive? Absolutely. Impossible? Not in the quantum world, where the wave nature of particles makes (almost) anything possible. The secret lies in the continuity of wave functions: they cannot suddenly just drop to zero.

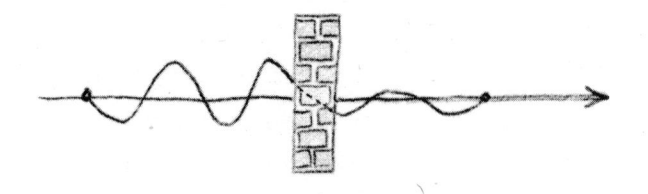

Quantum tunnelling

Instead, the amplitude of the wave function gradually decreases as it passes through the barrier. If the barrier is not too wide or too high, the wave function will emerge on the other side, and henceforth there is a chance that the particle is found there. As long as the curve hums along, the particle is everywhere and nowhere, surfing its quantum wave. The wider the barrier, the less likely the wave is to make it to the other side. Similarly, the probability of a particle passing through a wall decreases as the particle increases in mass. That's why marbles, for example, always bounce back wildly.

The first manifestations of tunnelling were discovered in chemistry. Friedrich Hund calculated the frequencies at which electrons are able to tunnel between the orbitals of various atoms. Robert Oppenheimer later calculated how long it takes for electrons to escape from hydrogen atoms when exposed to a uniform electric field (in this case, the electrons must surmount an energy barrier equal to the binding energy of the hydrogen atom). The most spectacular applications of tunnelling are found in the field of nuclear physics, a subject we will discuss in more detail in Chapter 7.

4.5 Matrix mechanics

> *Ecce est percipi.*
> *('To be is to be perceived.')*
> – GEORGE BERKELEY

Werner Heisenberg was born at a time when the quantum era was already underway. He was literally brought up on quantum. He was a student of Born, a confidant of Bohr and, like many of his peers, was well versed in a less algebraic form of art: the piano. After a violent hay fever attack, but more so to escape his teacher Bohr's relentless need for discussion, Heisenberg had fled to Germany's Heligoland, far away from polemics and pollen. There, in peace and quiet, he focused on what every physicist of the time was grappling with: developing a coherent quantum theory. Remarkably, Heisenberg's work arrived at the same conclusions as Schrödinger's – as if superposition had suddenly assumed human proportions. Yet the two men had started from entirely different questions, taken separate approaches, and used mathematical methods that couldn't have been more different. While Schrödinger built on de Broglie's ideas, Heisenberg forged his own path in Heligoland, where the sunsets seemed to stretch endlessly. Spoiler alert: Schrödinger's theory eventually became the more widely adopted of the two. It aligned more closely with traditional ways of thinking in physics and, perhaps most importantly, was easier to interpret.

Heisenberg had discovered a very specific (read: rather abstract and complex) mathematical approach to represent the position, momentum and energy (in sum: 'the state') of a particle. His method involved arranging numbers in rows and tables that could, theoretically, stretch on infinitely. He immediately realized the significance of his breakthrough: the wave/particle duality was unmistakably woven into his framework.

When Max Born reviewed Heisenberg's work, he made the link to Hamilton's work. He pointed out that Heisenberg's cryptic, crossword-like representations were actually simply matrices. Born and Heisenberg joined forces and the duo was further strengthened by Pascual Jordan (1902–1980). In this team of three, they diligently tinkered with Heisenberg's method. Their joint paper was published shortly after that same magical year of 1925, with the title *Matrizenmechanik*.

The essence of Heisenberg's method is that the position and momentum of a particle are represented by matrices that do not commute with each other – that is: changing their order alters their product. Multiplying momentum by position is not the same as multiplying position by momentum. This reveals a crucial insight: position and momentum are not independent variables. The main consequence of this is that you cannot possibly know the exact position and momentum of a particle at the same time. However, just because you cannot know them exactly does not mean that you cannot know them a *bit*. And that bit, in mathematical terms, is Planck's constant: the uncertainty of the position multiplied by the uncertainty of the momentum must exceed h (Planck's constant). The more precisely you measure a particle's velocity, the fuzzier its position becomes – and vice versa. Heisenberg's uncertainty principle is a cornerstone of quantum physics. You see: there are still certainties in life.

Essentially, you can never be quite sure of a particle's exact location. And this isn't due to flawed instruments or human error; it's a fundamental property of nature, which cannot simply be measured like that. The trajectory of atoms or electrons is completely different from the defined movements we usually imagine, in a clear line from point A to point B.[1]

There's more to it. Calculations can predict where a particle is most likely to be at the exact moment of measurement, but repeat the measurement later, and it might be somewhere completely different. As for where the particle resides between two measurements: we have no way of knowing. For Heisenberg, though, this was irrelevant. He cared only about what could be observed and measured. Where an electron might be (or not be) between two measurements was a question that should not even be asked.

Schrödinger realized that his theory, despite the different approach, was suspiciously similar to Heisenberg's theory. He didn't seem too happy about that – the two of them weren't exactly best mates. But what did the world care about personal grudges? The similarities between the two theories hadn't escaped Pauli's attention either, but it was Paul Dirac who pointed out that Heisenberg's matrix mechanics and Schrödinger's wave mechanics were fundamentally comparable. With some clever conversions and a tiny bit of tinkering, one theory could be perfectly derived from the other. The equivalence, however, wasn't immediately obvious; it was subtle and rooted in the many ways infinities can be represented. Schrödinger represented his wave functions as functions of a continuous variable x in a finite interval, while Heisenberg represented quantum systems in terms of matrices that had an infinite number

---

1 This is at least the case when it comes to very small particles. Large particles, with a large mass, move in an almost straight line. A key part of the construction of Heisenberg's theory was the condition that particles with a large mass or *quantum numbers* must behave like particles described by Newtonian physics. This is Bohr's correspondence principle.

of rows and columns. These representations are indeed equivalent. In both cases there is a one-to-one correspondence (provided by the Fourier analysis from Chapter 1) between natural numbers and continuous functions on a finite interval. The infiniteness of both representations is therefore the same. This is the beauty – and the power – of infinity. For eternity, amen.

4.6 Beauty is truth, truth is beauty

The twentieth century kicked off with a flourish of bells and whistles. The theory of relativity provided the bells, while quantum theory brought the whistles. Whereas the former focused on the speediest particles, quantum theory concerned itself with the tiniest ones. For years, the two theories lived side by side like estranged siblings, despite both inheriting their intellectual DNA from a shared grandfather: Albert Einstein.

Paul Dirac (1902–1984) was, like Einstein and John Keats, captivated by the elegance of mathematical formulas in physics: 'Beauty is truth, truth is beauty – that is all ye know on earth, and all ye need to know.' Dirac was a first-class loner, socially reserved, but a tireless worker, blessed with a very dry sense of humour; a true outlier in quantum land. In the summer of 1925, Heisenberg gave a lecture in Cambridge about his latest discovery, matrix mechanics, although at that time he had not yet realized how progressive his theory was – nor had anyone else. Through various channels, Heisenberg's transcript ended up in Dirac's hands. Dirac immediately recognized its deeper implications and set to work, zealously expanding on Heisenberg's ideas.

Paul Dirac: 'I do not see how a man can work at the frontiers of physics and write poetry at the same time. They are in opposition. In science you want to say something nobody knew before, in words which everyone can understand. In poetry you are bound to say something that everybody knows already in words that nobody can understand.'

It must be said that Dirac had a string of bad luck when it came to the timing of the publication of his papers. Based on Heisenberg's work, he developed an alternative description in terms of wave equations, but Schrödinger beat him to the punch by publishing his equation first. Dirac independently deduced the uncertainty principle, only to be overtaken by Heisenberg. Using Heisenberg's matrix mechanics, he calculated the spectrum of the hydrogen atom, but Pauli published his paper three days earlier. He had even discovered quantum field theory (see Chapter 7), but this time Jordan got the credit. And let's not forget Enrico Fermi, who simultaneously, but independently, discovered how quantum mechanics could describe a large number of electrons. Fortunately, the latter finding was also named after Dirac (Fermi–Dirac statistics).

Dirac's earlier efforts may have been mere 'warm-ups'. He made his real entrance through the main gate. Many of his discoveries were entirely his own. They mark him out as one of the greatest architects of quantum physics.

Heisenberg viewed quantum physics as a clean break from Newton's classical physics, seeing no meaningful connection between the two. What he overlooked was that there are different ways of doing classical physics (for the curious: through Hamiltonians, with Poisson brackets, or à la Lagrange). Dirac saw things differently. He uncovered profound parallels – many, in fact – between classical physics and Heisenberg's matrix mechanics. Dirac demonstrated that classical physics could, in fact, be quantized. He demonstrated that you can convert the equations of classical physics to equivalent descriptions in quantum physics in various ways (and how to do so). This was a landmark achievement, providing the foundation for understanding and describing many-particle systems in the quantum world.

PATHS AND INTEGRALS

The most tangible way in which Dirac converted classical equations to the quantum world was by using the path integral formulation. Initially a mere footnote in one of Dirac's papers, it was later embraced by Richard Feynman, with huge success. A quick note on path integrals. To describe the evolution of a particle from position A to position B, we draw all possible (meaning infinitely numerous) paths that a particle could take to get there.

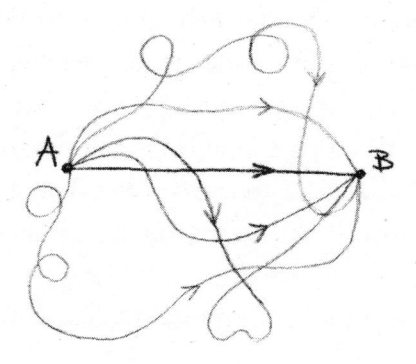

> Each of these paths is assigned a complex number: the action. The more twists and turns a path has, the larger this number becomes. According to classical physics, a particle chooses the path of least action. But in quantum physics, a particle takes all possible paths at the same time. In this case, the particle is smeared out over the multitude of paths. Using complex numbers, it becomes possible to calculate precisely the probability of a particle being somewhere at any given point. The fluctuations around the classical path depend on the particle's de Broglie wavelength: the smaller the wavelength, the more classical its behaviour, just like in the double-slit experiment. In a sense, the evolution of a particle is a type of double-slit experiment with an infinite number of slits. Points far from the classical path have an increasingly low probability of being traversed by a particle because, at these points, the wave function sums up countless positive and negative contributions, effectively cancelling each other out.

Nevertheless, Dirac is best known for something else entirely. This is the man who managed to reconcile Einstein's special theory of relativity with quantum theory. With his Dirac equation, he was able to describe particles that are both small (quantum) and fast (relativity). Time and space appeared to be more closely connected than anyone had imagined. That sounds like symmetry!

It's important to note that the arrival of quantum physics did not mean that Newtonian physics suddenly fell by the wayside. One is by no means a substitute for the other. Classical physics remains valid, provided we're dealing with large objects. In his trademark vague but correct way, Niels Bohr referred to this as the correspondence principle: in the realm of large quantum numbers, quantum physics reduces to classical physics. This correspondence principle played a key role in Heisenberg's matrix mechanics, because it allowed the calibration of observables (the properties of a system that can be measured). Whether you can use quantum or classical physics depends entirely on the de Broglie wavelengths. What counts as large or small is determined by Planck's constant. Incidentally, the same goes for the theory of relativity. Newtonian physics should not be discarded simply because relativity exists.

The theory of relativity, which deals with particles moving at very high speeds, reduces to classical theory when it comes to describing particles that move slowly. Whether something is considered fast or slow is determined entirely by its relation to that other constant: the speed of light.

And while we are at it: just because Dirac had formulated a relativistic version of Schrödinger and Heisenberg's theories does not mean that the latter were on the wrong track. As long as you are describing particles that move slowly in comparison to the speed of light, Schrödinger's approach serves just as well.

In summary:
— Large + slow = Newton (classical theory)
— Large + fast = Einstein (theory of relativity)
— Small + slow = Schrödinger/Heisenberg (wave equations)
— Small + fast = Dirac (wave equations)

If you use the Dirac equation to describe particles moving much slower than the speed of light, you end up with the Schrödinger equation. But that's not entirely true: you get the Schrödinger equation with an extra twist. That twist entails an extra degree of freedom.

An additional property of the electron is introduced: spin. For electrons, the spin has a value of ½ (one half), meaning that a measurement can only give two different outcomes: 'up' or 'down'. This is the subject of the Stern–Gerlach experiment – more on this in a moment – which revealed that the spin behaves like a tiny magnet. Although spin is a very abstract concept, the spin movement can be loosely compared to a spinning top turning to the right (up) or to the left (down). The spin has the value ½ because, oddly enough, it takes a full 720-degree rotation – not the usual 360 degrees – to bring it back to its original state.

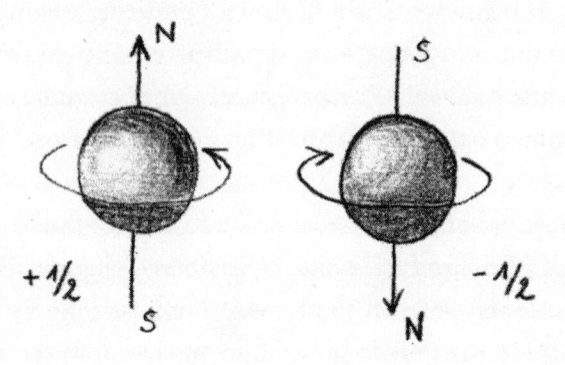

How an electron spins. Here respectively up, $|\uparrow\rangle$, and down, $|\downarrow\rangle$.

Dirac's prediction of the existence of a spin ½ corresponded exactly to a trick Pauli had introduced to explain Mendeleev's periodic table on the basis of his exclusion principle. Pauli had introduced a spin with a well-defined magnetic momentum that could assume exactly two different values. This property explained, among other things, how two electrons could occupy the same energy orbital. Pauli's trick worked, but Dirac was the first to understand *why* it worked.

From this point onward, everything in quantum physics builds not only on Schrödinger's equation, but also on Dirac's. That alone is an extraordinary achievement. But before diving into the implications of his work, let's add one more feather to Dirac's already impressive cap. After a lot of thinking and calculating with his equation, Dirac predicted that, if there exists an electron with a negative charge, there should also exist an antielectron with a positive charge. And as a matter of fact: this antiparticle was discovered in 1932 and given the less-than-imaginative name 'positron'. This gave Dirac's name even more weight, because by discovering the positron he realized the dream of every theoretical physicist: to make a daring prediction that is later spectacularly confirmed by an experiment, however crazy or difficult.

Dirac also introduced the term 'quantum vacuum', a reference to the one true and absolute nothingness – a descendant of the black box family. Dirac stated that even supposedly 'empty' systems are teeming with life. At absolute zero, the void is a vast and complicated soup of quantum particles, collectively known as the 'Dirac Sea'. This sea encompasses all particles with a negative energy. Each additional negative-energy particle that is added to the Dirac Sea lowers the total energy, as nature relentlessly pursues the lowest possible energy state. To make sense of this, imagine dropping Emmy Noether's goldfish into this large Dirac Sea. Through its microscopic eyes, the goldfish would perceive nothing unusual; everything would seem like water. The only visible features would be the places where there is really nothing. A bubble rising to the surface, for example, because that bubble has a positive energy. In this flipped perspective, positive and negative are simply reversed: all the goldfish sees is nothingness. And it can only see places where there really is nothing. This is precisely what quantum physics grapples with: the nature of absolute nothingness. That may sound like particularly bad advertising, but in truth, it's the best publicity for the sciences! After all, anyone can be concerned with anything and everything, and anyone can understand something. But to understand nothing, and by doing so, comprehend everything?! Who could possibly resist?

Dirac's formalism went so far mathematically and took on such abstract forms that it appeared to lose all touch with reality – and with physics itself. It was difficult to feel any connection to it. Intuition was gradually sidelined, turning quantum physics into a purely mathematical matter, verging on the mysterious. 'I think,' concluded Richard Feynman pragmatically, 'I can safely say that nobody understands quantum mechanics.'

THE SMARTEST PERSON IN THE WORLD

Whereas Schrödinger, Heisenberg and Dirac handled their mathematics 'like engineers' (i.e. a bit sloppily – they simply wanted things to work), John von Neumann (1903–1957) meticulously redeveloped all of their formulas. This Hungarian-American mathematical powerhouse was unmistakably the brightest mind of his time: a cheerful *bon vivant* you would definitely want at your birthday party, who liked to read at the wheel and who, as a teenager, effortlessly memorized a twenty-volume encyclopaedia. Decades later, he could still recite what was on page 1,729 of volume seven. Von Neumann also invented the basic architecture of the modern computer. Though there are 1,001 other formidable anecdotes about this man, let's stick to the quantum essentials.

Von Neumann ultimately made all quantum theories clear, consistent and free of contradictions. He rigorously proved that Schrödinger's and Heisenberg's representations of quantum mechanics are equivalent (for the aficionados: all ways of representing the canonical commutation relations are equivalent; this is the content of the famous Stone–von Neumann theorem). He provided quantum physics with a solid mathematical basis

> by formalizing it within Hilbert space, an infinitely large space that houses all possible wave functions. A problem with Hilbert space is that it has an infinite number of dimensions. It took a genius like von Neumann to see the wood for the infinite trees. But thanks to him, Hilbert space has become the favoured playground for entire generations of physics students.

4.7 The spin becomes a qubit

STERN AND GERLACH'S EXPERIMENT

In 1922, Otto Stern and Walther Gerlach carried out an experiment, which led recalcitrant physicists who were as-yet unconvinced of quantum theory to reconsider their views. This now-famous experiment demonstrated that the spin of an atom is undeniably quantized (able to take on only two values when measured) and could only be explained using quantum physics. Einstein, recognizing immediately its significance, promptly nominated both men for the Nobel Prize.

But first: who were Stern and Gerlach? Otto Stern (1888–1969) was Einstein's first pupil and, in keeping with the saying, 'If you lie down with dogs, you will get up with fleas', he caught Einstein's fascination with light quanta, atoms and electromagnetism. But Stern clearly also inherited his teacher's scepticism and rejected the model of an atom proposed by Bohr, who had previously predicted that the magnetic momentum of an atom is quantized. If Bohr's drivel was true, Stern claimed, then he'd be done with physics. Determined to disprove Bohr, he devised an experiment to prove the opposite: that spin is *not* quantized.

Walther Gerlach (1889–1979) was the intrepid experimentalist who put together the Stern–Gerlach experiment. Upon completion of the experiment, Pauli sent his congratulations by post – though not, of course, without a phlegmatic sharp edge: 'Hopefully now even the incredulous Stern will be convinced about directional quantization.'

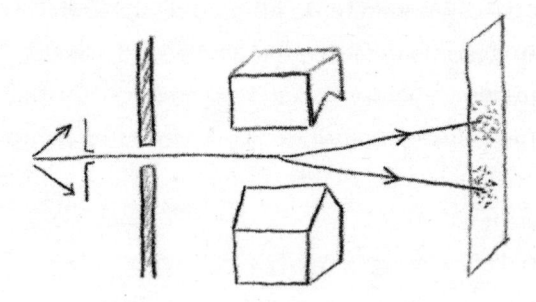

The Stern–Gerlach experiment. Silver atoms are beamed from a source. The north and south poles of the magnet deflect their trajectories.

Stern and Gerlach didn't need a lot for their experiment: a magnet, a projection screen and a few million silver atoms. One by one, the silver atoms were fired through the vertically positioned magnet (positive pole above, negative pole below). Stern, adhering to the classical assumption that the magnetic momentum of atoms could point in an arbitrary direction, expected the atoms to be deflected in a correspondingly random fashion by the magnetic field, eventually forming a single extended vertical beam on the screen. After all, the more aligned the spin was with the magnetic field, the greater the deflection should be.

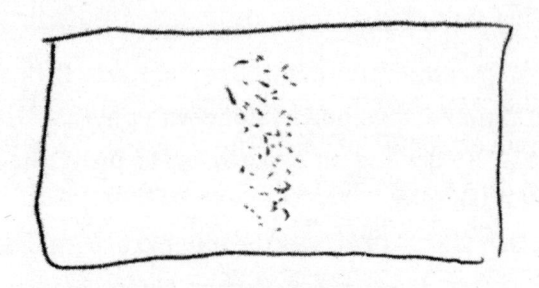

What they expected to see.

What did they see? Something completely different! Instead of a single elongated streak, two sharply defined, distinct vertical 'clouds' of atoms appeared on the screen. The result delivered a message with a silver lining: the spin can only assume two values. Therefore, the spin is quantized. Additional evidence was provided by the fact that when the magnet was rotated by a certain angle, the two 'clouds' rotated with the same angle. Astonishment all around: the spin is quantized in all directions!

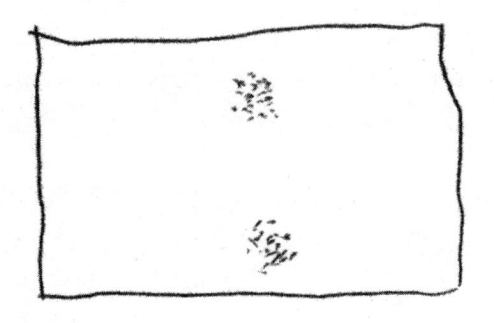

What they saw.

Everyone – even Einstein – agreed on one thing: this experiment could not possibly be explained using classical physics. So, who (or what) was behind it? It took another three years for all the pieces of the puzzle to fall into place. Incidentally, this experiment owes its success to a rather funny twist of fate. Atoms are extraordinarily small – practically invisible. Stern and Gerlach learned this the hard way when, at first, they saw no results at all. It just so happened that Stern, an avid cigar smoker, could not afford the best tobacco on his meagre assistant's wage. This turned out to be a blessing in disguise, as cheap cigars contain a lot of sulphur. The sulphur smoke from the cigar unexpectedly reacted with the silver on the projection screen, turning it into jet-black silver sulphide. Suddenly, the previously elusive result of their experiment appeared as two distinct clouds. Gerlach promptly lit a cigar of his own.

How did the puzzle fit together? Silver atoms have a spin. As a result, they behave like very small magnets that can only assume two values: up or down (and any superposition thereof). If a particle passes through an external magnetic field, an upward or downward force is exerted on it, depending on whether its spin is up or down. But if the particle's spin is in a superposition of up *and* down, the particle itself splits. The part with positive spin is pushed upwards (eventually landing at the top of the screen) while the part with the negative spin is deflected downwards (landing at the bottom of the screen). In this case, the distance between the two paths can far exceed the de Broglie wavelength of the silver atoms. The path the particle takes becomes entangled with its spin. By hitting the screen (the observation), the wave function breaks (the 'collapse'), resolving the superposition into a definitive state. It is no longer both/and, but either/or. Either the particle ends up in one cloud, or in the other – but never in both at the same time.

Why was this experiment so important? Because it confirms a number of crucial insights. First of all, it validated the superposition principle. Whereas in classical physics a particle is either in one state or another, particles can, from a quantum mechanical perspective, be in two states simultaneously, at a distance that is much greater than their de Broglie wavelength. As mysterious as it may be, superposition is the quintessence of quantum. The second doubt dispelled by this experiment: a spin is quantized. It can only have two directions: spin-up or spin-down (along with any superposition of these states).

And because good things come in threes, this experiment had even more to reveal. What if we arranged several magnets in a row? And what if we were to rotate certain magnets ninety degrees? And what if we were to block certain paths while leaving others open? By pouring all these questions into the experiment, a third very important conclusion was reached: interference really does exist. Atoms can not only exist in more than one place at the same time, their properties can also mutually influence each other (think of the waves that can reinforce or attenuate each other). This is, sure enough, the hardcore version of the double-slit experiment! Lastly, the experiment solidified one final, profound truth: a measurement does have an effect on a (quantum) system.

THE QUBIT

Bearing Einstein's motto in mind that everything should be made as simple as possible, but no simpler, physics students today are introduced to quantum physics via Stern and Gerlach's experiment. And this inevitably requires knowledge of the concept of the qubit, which gets its name from 'quantum bit'. The following paragraphs illustrate why quantum physics can be considered one of their most difficult subjects.

In essence, the qubit is a compressed version of Schrödinger's wave functions. While Schrödinger's wave function allows a particle to exist in infinitely many locations simultaneously, the selection menu for a qubit consists of only two possibilities: it's either here or it's there. Similarly, Heisenberg's Hilbert space consists of an infinite number of vectors ($|0\rangle$, $|1\rangle$, $|2\rangle$, $|3\rangle$, etc.), whereas for the qubit the choice is limited to $|0\rangle$ or $|1\rangle$. But the superposition principle dictates that it can be $|0\rangle$ and $|1\rangle$ at the same time.

For comparison, classical computers operate on the basis of two units: zeros and ones. These are the bits. A sequence of bits can occur in exponentially many combinations, representing numbers and information.[1] A transistor does nothing other than convert zeros and ones into other zeros and ones.

Bits are not computer-bound. They can also take forms other than 0 or 1. A switch, for example, has two positions: 'on' or 'off'. The lights on your Christmas tree flash on and off. Or a (classical) relationship: it is either on or off.

Qubits, like bits, are an abstraction. They represent any quantum system that can exist in two different states – or any possible superposition of those states. An electron, for example, can have a spin that is both 'up' and 'down' at the same time. Another example is the polarization of light, which can be right-handed and/or left-handed. Or non-classical relationships, which are both 'on' and 'off', but there is mainly a lot of ambiguity. Electron orbitals provide yet another illustration: an electron can partially occupy one orbital (the S orbital) while simultaneously occupying another (the P orbital). What makes quantum physics so difficult – or so fun – is that you have to figure out exactly how much an electron is in one orbital and how much in the other.

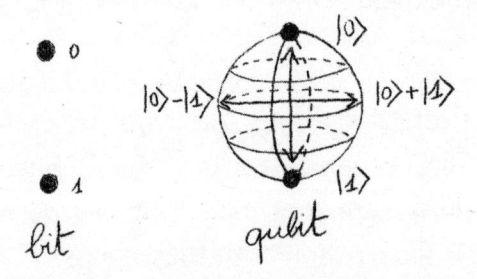

A classical bit, on the left, is either unequivocally 0 or 1. A qubit (on the right) can be 0 or 1, but it can also be in a superposition of 0 and 1. A qubit can assume any possible value on a sphere. Four possible values are indicated on the illustration: |0⟩, |1⟩, |0⟩ - |1⟩ and |0⟩ + |1⟩.

---

[1] This is why the decimal system, popularized by – again – Simon Stevin, is so much more powerful than the Roman numeral system, in which the number of combinations does not increase exponentially with the number of symbols used.

To visualize this, imagine a perfectly round globe. Each point on its surface corresponds to a possible quantum state of a qubit. The state |0⟩ corresponds to the north pole, while |1⟩ is at the south pole. A spin pointing in any other direction can always be written as a superposition of |0⟩ and |1⟩. For instance, two particular states on the equator are |0⟩ +|1⟩ and |0⟩ - |1⟩. These are qubits polarized in the x-direction, and can be regarded as the qubit version of a state with a specific momentum. If you measure a qubit's position, the result will be either |0⟩ or |1⟩; if you measure its 'momentum,' the outcome will be |0⟩ + |1⟩ or |0⟩ - |1⟩. A qubit cannot have an unambiguous position *and* momentum at the same time, which is fully compatible with Heisenberg's uncertainty principle. Any two opposite points on the globe correspond to orthogonal states, forming a basis in which measurements can be made.

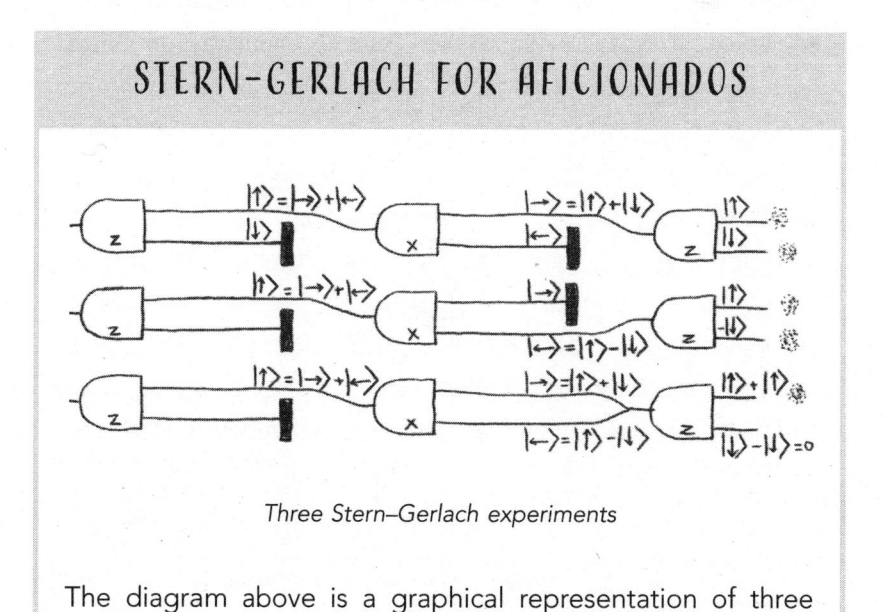

STERN–GERLACH FOR AFICIONADOS

Three Stern–Gerlach experiments

The diagram above is a graphical representation of three different Stern–Gerlach experiments. Silver atoms are emitted

THE FIRST QUANTUM REVOLUTION 135

from the source (far left) and their spin is in an arbitrary superposition of $|\uparrow\rangle$ and $|\downarrow\rangle$. This is represented as: $|\psi\rangle = a|\uparrow\rangle + b|\downarrow\rangle$. When these silver atoms pass through a vertical magnetic field (z), the wave function splits into two paths: the spin-up part takes the upper path and the spin-down part takes the lower path. By blocking the lower path and only allowing spins from the upper path to pass through, we have found a way of isolating the spin-up atoms: $|\psi\rangle = |\uparrow\rangle$.

Next, we send these spin-up atoms through a magnetic field oriented in the x-direction (x). This causes the atom to be deflected either to the left or to the right. But how can we know the x-direction of the spin (left or right?) if all we know is that the particle has a spin-up? You can write down a spin-up in the basis of the spins in the x-direction, resulting in a superposition: $|\uparrow\rangle = |\rightarrow\rangle + |\leftarrow\rangle$. This means that a spin-up is the superposition of the left *and* right spin! For the spin-down, this gives $|\downarrow\rangle = |\rightarrow\rangle - |\leftarrow\rangle$ (note the sign difference). But equally, we can express the left and right spins in terms of spin-up and spin-down: $|\rightarrow\rangle = |\uparrow\rangle + |\downarrow\rangle$ and $|\leftarrow\rangle = |\uparrow\rangle - |\downarrow\rangle$. In other words, the spin can be described both in the z-axis and in the x-axis. These are two complementary ways of describing a spin, analogous to Heisenberg's complementary variables of position and momentum. After passing through the x-magnet, the wave function splits into a left and a right path. Now, let's examine the outcomes of three different setups. By blocking the left path, only right-spin atoms are allowed through: $|\psi\rangle = |\rightarrow\rangle = |\uparrow\rangle + |\downarrow\rangle$.

When these atoms pass through another z-magnet, the spin-up and spin-down will follow different paths, forming two distinct clouds on the screen.

In the second setup, we block the upper path, leaving only

> spin-left atoms. Since $|\leftarrow\rangle = |\uparrow\rangle - |\downarrow\rangle$, these atoms will split up again when passing through the z-magnet into $|\uparrow\rangle$ and $-|\downarrow\rangle$. Since the minus sign has no effect on the measurement, we get exactly the same result as in the first experiment: two clouds appear on the screen.
>
> However, the third setup is much more interesting. Here, neither path is blocked. A superposition of spin-left and spin-right atoms is sent through the z-magnet. In this case, the spin-down paths will negatively interfere with each other (cancelling each other out): $|\rightarrow\rangle + |\leftarrow\rangle = (|\uparrow\rangle + |\downarrow\rangle) + (|\uparrow\rangle - |\downarrow\rangle) = |\uparrow\rangle$. Result: only one cloud is detected, namely the upper one. The probability of the atom ending up in the lower cloud is zero. This is interference *par excellence*. The third experiment is clearly not the sum of the previous two.

It might surprise you, but qubits are nowhere near as abstract as you might think. Your body, which is largely water (H_2O), is essentially a vast collection of qubits. The nucleus of each hydrogen atom (a proton) has a spin that can be represented as a qubit. When your body is placed in an MRI (Magnetic Resonance Imaging) scanner, these proton spins interact with a powerful magnetic field. The scan detects the distribution and density of water in your body, allowing it to infer the type and location of different tissues. By applying a strong magnetic field, most of the proton spins align with the field, settling into the state $|0\rangle$. A radio wave then excites these spins into a superposition state, $|0\rangle + |1\rangle$. Over time, the spins naturally return to their lower-energy state, releasing energy. It's that released energy that is measured and captured on the scan, creating a detailed map of your body.

IN A NUTSHELL

> Quantum physics describes the physical world. Metaphysics deals with the spiritual world. Quantum physics has nothing sensible to say about metaphysics.

> Quantum physics is inherently non-local: entangled particles, whether far apart or close together, can influence each other.

> Einstein doesn't believe in randomness. Nor in non-locality. But he does believe in local realism. He claims he can prove he is right, using the EPR thought experiment. Goodbye to quantum physics, says Einstein. Bohr counters. Schrödinger sends his cat.

> John Bell devises an experiment to give a definitive answer on local realism. Bingo.

> Quantum physics is contextual: the properties of a system are not absolute. They vary depending on how you look at it.

> *Dramatis personae*: Ann Tanglement, Albert Einstein, Niels Bohr, Erwin Schrödinger, John Bell, Grete Hermann.

FIVE

QUANTUM PHILOSOPHY

5.1 Quantum spiritualism

> *There are only two ways to live your life. One is as though nothing is a miracle. The other is as though everything is a miracle.*
>
> – ALBERT EINSTEIN

Quantum physics is not philosophy. There is of course a philosophical side to it – a very important part, even – so it needs to be addressed, sooner or later. But it's not the essence. The aim of quantum physics is to predict the outcome of a measurement or to determine how a system of quantum particles (such as an electron in a metal, in a molecule, in a transistor) will respond to external impulses. And in that sense it works perfectly well, despite its extreme uncertainty and unpredictability. Just about every scientific success in quantum physics is down to the very Feynmanian 'shut up and calculate' attitude, founded on theories, experiments and predictions. Do the experiments align with the predictions? Great, let's keep going! Do they contradict them? Fine, start over – but keep going anyway. Without a thorough, rational understanding of electrons, light and atoms, humanity would never have achieved such quantum leaps in the past 120 years. Nor would quantum physics have embedded itself so indispensably into the technologies that shape our daily lives.

The point we want to make here is that quantum isn't magic. Just because something has various confusing or bizarre properties doesn't mean it was conjured out of a magician's hat. Fascinating and counterintuitive? Absolutely. But magical? Not at all. Our brains simply aren't designed to grasp everything. And yet these 'spooky properties' are precisely the reason why the applicability of quantum physics is often misinterpreted; they create the illusion that quantum physics should provide a scientific explanation for alternative therapies and telepathy, quantum healing and spherical vortices. For those who believe in paranormal phenomena, it is indeed a quite natural – and logical – reflex to assume that quantum physics provides a scientific explanation. After all, isn't quantum physics about energy levels, things influencing each other at a distance, vibrations and resonances? Well, yes. But as far as we know, human consciousness is still not part of the grand quantum energy field from which some claim one can extract free energy – provided you're willing to pay for it. Quantum physics cannot be applied to our experiences in everyday life. When we talk about elementary particles becoming entangled, we mean atoms, electrons and their kin – not someone's consciousness mysteriously intertwined with the energy state of a right-spinning co(s)mic sphere floating in a cuddly alternate universe. Which leads us seamlessly to our chapter on quantum entanglements, a tale of how quantum physics gradually, albeit in fits and starts, befuddled the great minds of the twentieth century.

5.2 Entanglements

Bohr was inconsistent, unclear, wilfully obscure, and right.
Einstein was consistent, clear, down-to-earth, and wrong.
— JOHN BELL

Entanglement is one of the most important and radical consequences of the principle of superposition in quantum physics. It refers to the correlation that continues to exist between two particles that once interacted but at some point became separated. Since these particles initially formed a single, unified entity, their properties should still be viewed as such, since they are always part of the same wave. But what does it actually mean, when two particles are entangled?

First: if you perform a measurement on one particle, this has an immediate effect on the properties of the other, no matter how far apart they are – be it one particle in Anatolia and the other in Zanzibar. In the case of a 'maximally entangled Bell state', if the measurement result for particle 1 is spin-up you automatically know that particle 2 is spin-down (and vice versa). This phenomenon, known as 'non-locality' (or 'action-at-a-distance') is pretty much the most peculiar aspect of quantum theory – and was a persistent thorn in Einstein's side.

Second: if you measure and observe only one of the entangled particles, you get a completely random outcome. This is because the real information is entirely enclosed in the connection (the 'correlations') *between* the results of measurements on both entangled particles. John Preskill (b. 1953) illustrates this very aptly during seminars using the example of a quantum book: in a traditional book, you can read one page and then another, perhaps skip a few pages, and even by skimming through parts, you'll eventually grasp the story and live happily ever after. A quantum book is nothing like this. In a quantum book, each individual page is completely meaningless on its own. You only get the full story (the information) by analysing the connections between *all* of the pages. Remove even a single page, and all the information is lost for ever. The essence lies in the correlations.

5.3 Niels versus Albert

> *I have tried to read philosophers of all ages and have found many illuminating ideas but no steady progress toward deeper knowledge and understanding. Science, however, gives me the feeling of steady progress: I am convinced that theoretical physics is actual philosophy. It has revolutionized fundamental concepts, e.g., about space and time (relativity), about causality (quantum theory), and about substance and matter (atomistics), and it has taught us new methods of thinking (complementarity) which are applicable far beyond physics.*
>
> — MAX BORN

There were certain things that Einstein was firmly convinced of. One: *Zufall* (German for 'chance') does not exist. Two: since *Zufall* does not exist, it also cannot be an inherent property of nature. Three: things do not change their properties just because you look at them. For Einstein, the chance factor in quantum experiments was nothing more than a logical consequence of our lack of knowledge and the existence of other (non-observable) particles. Einstein firmly believed in the existence of 'hidden variables', an invisible 'clockwork universe' in which nature operates in a completely deterministic way. And what better way to study the chance factor than by scrutinizing the uncertainty principle? The uncertainty principle compels us to ask different, more philosophical questions. Does an object have properties depending on whether or not you look at it? Do those properties change depending on the experiment you perform on it? And can nature 'know' when it's being measured?

Challenging the uncertainty principle became a passion project for Einstein, almost a hobby like no other. He was determined to prove that the entire quantum theory, as introduced by Schrödinger and Heisenberg in 1925, was perhaps correct but was not complete. A theory that relies on wave functions and predicts instantaneous effects across arbitrarily long distances could not possibly offer a full description of reality. On this, Schrödinger was in complete agreement. Both men struggled to come to terms with the interpretation of the latest quantum theories, despite being largely responsible for their creation.

In the end, Einstein let things be. He was confident that, in time, others would come to realize that quantum theory was nothing more than a stepping stone to a more accurate theory, one in which the hidden variables did have a rightful place and the 'spooky action-at-a-distance' did not: an all-encompassing classical theory that would unite quantum theory, gravity, magnetism and electricity.

The perfect stage for Einstein to launch his (well-meant) attacks came in 1927, at the fifth Solvay Conference. The Solvay Conference is like Beethoven's Fifth Symphony: the most iconic and most played of its kind. Since its inception in 1911, held every three years at the Hotel Métropole in Brussels, the Solvay Conference has brought together the brightest minds in chemistry and physics to tackle the most pressing questions of the time. The 1927 conference, held in the wake of the revolutionary 1925 developments, was dedicated to the brand-new quantum theory. Presiding over the meeting were Niels Bohr and Albert Einstein, respectively the leading advocate and the most eloquent opponent of quantum physics. The guest list read like a who's who of modern physics, including Erwin Schrödinger, Wolfgang Pauli, Werner Heisenberg, Paul Dirac, Louis de Broglie, Max Planck and Marie Curie. Of the twenty-nine attendees, no fewer than seventeen were past or future Nobel laureates.

Unlike Heisenberg, who at the time literally ran away from the towering presence of Bohr (whom he sometimes thought too much a philosopher and not enough a physicist), Einstein could have wished for no better antagonist than Bohr in his *offensifitis* against quantum physics. The titanic struggle between the two went something like this: at night, Einstein would lie awake, tirelessly concocting thought experiments designed to prove that Heisenberg's uncertainty principle was full of holes. And as mentioned before, Einstein had an unparalleled knack for devising thought experiments. In the morning, he would invariably spoil Bohr's breakfast by dishing up yet another argument as to why quantum theory was wrong, or at least incomplete. Bohr, undeterred, would then spend the whole day brooding over counterarguments. In the evening, the two met again over dinner. And by the time the plum pudding was served, Bohr would triumphantly dismantle Einstein's thought experiment, showing that, no matter how clever or innovative, it ultimately failed to refute the uncertainty principle. Thus, Einstein would plunge into another sleepless night, regirding his loins, searching for his next counterargument in this clash of the titans.

EINSTEIN AT HIS SHARPEST, BOHR AT HIS BEST

As a brief, though not completely unrelated tangent, let's delve into one of Einstein's most astute thought experiments. Central to this experiment is the uncertainty relation between energy and time, a variant of Heisenberg's 'ordinary' uncertainty relation between position and momentum. This energy–time uncertainty relation is a consequence of the properties of a wave packet: the precise time it takes for a packet to hit a detector depends on its size. The smaller the packet, the more different frequencies (and hence energies) it contains (another consequence of Fourier analysis). So the more precisely the time is defined, the less certain the energy becomes, and vice versa.

Simplified, Einstein's thought experiment went like this: a box is designed to open at a precisely defined moment. The box opens and closes again so quickly that only a single photon can escape. The box is weighed before the photon has escaped, and again afterwards. Since the box is connected to a clock, the exact moment of its opening is known.

By comparing the difference in weight before and after the photon escapes, it is possible to know both the change in energy and the precise moment this change occurs. This seemed to violate the energy–time uncertainty principle. Ha! Game over – or so Einstein thought . . .

While enormously impressed, Bohr had to sneakily call in reinforcements this time (led by Heisenberg and Pauli) to find the weak link in Einstein's reasoning. A few days later, Einstein was to get a taste of his own medicine. The master of the thought experiment had indeed overlooked a critical detail: the theory of general relativity. The latter dictates that the stronger the gravitational field (i.e. the closer to the Earth's surface), the slower time flows. Since the box is placed on a scale, with every photon that flies away the box will become lighter by a tiny but not insignificant amount, causing the scale to rise by a tiny but not insignificant amount. However, this tiny but not insignificant amount is just enough to restore the uncertainty relation between energy and time, since the difference in gravity has an effect on the clock. Satisfied with his rebuttal, Bohr quietly savoured his dessert. Re-energized, Einstein left the table to tackle yet another long night of reflection.

5.4 The EPR paradox

One morning in 1935, a full eight years after the famous 1927 Solvay Conference, Einstein arrived at the breakfast table of his Princeton home in a remarkably cheerful and triumphant mood. The coffee wasn't too hot, his egg was perfectly cooked and, notwithstanding all the excessive fanfare at the time around the opening of the Moscow metro (the first in the Soviet Union!), this was to be *his* day. He had just finished a new paper that he believed would finally deliver the fatal stab to the heart of quantum theory. Einstein duly submitted it to *Physical Review*. A week before the paper was published, his collaborator Boris Podolsky leaked the findings to the 'secular' press without Einstein's knowledge – and much to his annoyance. In Einstein's view, scientific matters belonged in scientific journals. Relations between Einstein and Podolsky would never recover.

As was common with everything Einstein did or didn't do in those years, the newspapers were again bursting with headlines, interviews, and other assorted titbits for the hungry public. Einstein was hot. Perhaps no one else either before or since has ever had such fame. Once again, he graced the front page of *The New York Times*. After all, delivering a death blow to quantum theory was, naturally, front-page material. The title of the paper in *Physical Review* was as striking as the controversy it sparked: 'Can [the] quantum-mechanical description of physical reality be considered complete?'

The New York Times.

NEW YORK, SATURDAY, MAY 4 1935 — TWO CENTS

EINSTEIN ATTACKS QUANTUM THEORY

Scientist and Two Colleagues Find It Is Not 'Complete' Even Though 'Correct.'

Quantum physics in The New York Times.

Einstein, together with fellow physicists Boris Podolsky and Nathan Rosen, had devised a thought experiment (generally known as the EPR paradox, from the initials of the authors' surnames) designed to show that the concept of quantum non-locality was wrong. He refused to believe that two particles, no matter how far apart, could instantaneously influence each other – or that a measurement performed on one entangled particle could have an immediate effect on the other. Because that would mean that the information concerning the change of state would have to flit from particle A to particle B at a speed beyond imagination, say instantaneously. Faster even than light. But ever since the theory of relativity (from the very same Einstein), no physicist doubts the dogma that nothing can travel faster than light. Einstein felt comfortably in the lead here. Entangled non-locality with its spooky action-at-a-distance and what-not; why bother him with that? It simply couldn't be right. The quantum theory didn't hold together. For him it was simple: two particles, two realities. Against non-locality, he countered with local realism, the idea that the properties of particles exist independently and are not influenced by distant events.

The EPR thought experiment runs like this: two quantum particles (1 and 2) are entangled. Their wave function has a peculiar property: if you know the momentum of one particle, you automatically know the momentum of the other. Ditto for their positions, where the particles are at distance L from each other. That distance needs to be significant because, in the time taken to perform a measurement on particle 1, the information from that measurement must not be able to reach particle 2. So, particle 2 cannot 'know' that a measurement has been done on particle 1. In this entangled state, neither the momentum nor the position of the particles is defined. For the aficionados, this is the wave function: $\psi = \delta(x_1 - x_2 + L)\,\delta(p_1 + p_2)$.

Put simply, particles 1 and 2 are in a superposition. They can be anywhere in space, but always at the same distance (L) from each other. Their speed can also vary, albeit in such a way that the two particles are always travelling in opposite directions ($p_1 = -p_2$). Now suppose a measurement is performed of the position of the first particle. The superposition ends, the wave function collapses (with plenty of dramatic flair) and the exact location of the first particle is defined precisely. Let's call this position x_1. At that moment, the position of the second particle is also instantly fixed, namely at locus $x_2 = x_1 + L$. With a different measurement setup, we could have measured not the position but the speed of the first particle. This would have resulted in a measurement of p_1. In that case the speed of the second particle would also have been instantly fixed ($p_2 = -p_1$). *Also sprach die Quantenphysik.*

What is Einstein's take on this? In Einstein's worldview, consistent with local realism, the type of measurement you choose to perform on the first particle (either position or speed) cannot possibly affect the second particle. Systems that are spatially separated from each other have their own reality, Einstein argued; their properties are very clearly defined in advance and exist independent of our observation or of any measurements. Therefore, he reasoned, if a measurement on particle 1 allows you to determine particle 2's position precisely, and a different measurement on particle 1 allows you to determine particle 2's momentum exactly, then both the position and momentum of particle 2 must have been precisely determined all along. Take that, Werner! After all, Heisenberg's uncertainty principle posits that position and momentum can never be known with precision at the same time. Einstein's conclusion: quantum physics is inconsistent – and Miss Ann Tanglement is a sneaky saboteur!

That morning at the breakfast table, Bohr too was reading the newspaper. He hadn't seen this storm coming. The whole quantum world was in uproar. What could this possibly mean? Einstein's paper made it seem that quantum theory, the uncertainty principle included, failed across the board. What would Pauli, 'the conscience of physics', think about it? That it was a disaster, a catastrophe! That's how Pauli saw it. Had all the work been for nothing?

As the lord and master of quantum physics, it fell to Bohr to respond. His thought processes promptly shifted into high gear.

After six weeks of mulling and pondering, he was finally able to take the sting out of Einstein's waspish game. The problem was that, at first glance, everything seemed to make sense. And yet, on closer inspection, something didn't quite add up . . . Only you had to look very carefully to find it. What it came down to was this: there *was* no paradox. Summarizing Bohr's counterargument is no easy feat. In true Bohrian fashion, it was wrapped in convoluted sentence structures and arcane reasoning. In any case, he delicately pointed out to Einstein that there is still such a thing as 'counterfactual reasoning', a concept almost as unintuitive as the phrase itself. Bohr reminded Einstein that it is not possible to measure position and speed simultaneously on the same system (particle 1), since these observables do not commute. As long as no measurement is performed on one particle, the other particle remains in a superposition. Only when you measure one particle do the properties of the second particle become fixed. So there is no way for the properties of the second particle to have been determined in advance. The only thing established, Bohr explained, is that if the position of particle 1 is measured, particle 2 will be found at a distance L from it, and if the momentum of particle 1 is measured, the momenta of the two particles will be opposite. That's all we know. The properties of a quantum system depend entirely on how you look at it. Position and momentum are complementary ways of describing the same quantum system, but they cannot be measured simultaneously.

The opposing side's conclusion was clear: the EPR paper in no way contradicts Heisenberg's uncertainty principle. Einstein had simply made a (very subtle) error in reasoning. He had applied the right mathematics, but not interpreted it properly. Quantum physics does indeed have the spooky property that a measurement on one entangled particle has an instantaneous effect on its partner –

because two particles that were once part of the same system remain entangled. Conclusion: Einstein's local realism is fundamentally incompatible with quantum physics. It's one or the other. But not both. Hurray for non-locality!

But hold on a nanosecond . . . What about that instantaneous communication implied by the EPR experiment? Didn't that contradict Einstein's theory of relativity? No, because no information is being exchanged here. The theory of relativity only says that information cannot travel faster than light. But Bohr and Einstein didn't fully grasp this distinction yet, since the mathematical concept of information is very subtle (and wasn't introduced by Claude Shannon until 1948). Information is not the same as correlation. Measurements on entangled particles cannot be used to send information from one particle to the other. This is because the measurement results on the first particle are completely random and randomness cannot be manipulated. Without control over the result, information cannot be transmitted. Conclusion: non-local correlations in quantum mechanics do not violate the theory of relativity.

With their thought experiment, Einstein, Podolsky and Rosen had triggered a full-blown philosophical discussion, whether they meant to or not. To be or not to be? To influence or to disrupt? Correct but not complete? Local realism (a consequence of a philosophical approach) or non-locality (a mathematical consequence of the axioms of quantum physics)? Was there really no one who could come up with a genuine experiment to resolve these fundamental questions once and for all? As it turns out, there was! Thirty years later, John Bell would rise to the challenge and deliver the answer. Despite sowing widespread confusion and discord, with no one knowing how to interpret what, the questions arising from the EPR paper did eventually ensure that the mystery called 'entanglement' would be unravelled.

5.5 Schrödinger sends his cat

Schrödinger had watched with relish the tug-of-war between Einstein and Bohr (or rather, between Einstein and most others). But unlike most others, he took Einstein's side. In a letter, he congratulated Einstein on the EPR paper, writing: '... it works as well as a pike in a goldfish pond and has stirred everyone up.'

Schrödinger, who was equally unconvinced by the way that Bohr and co. were interpreting (his own) quantum theory, agreed with Einstein's reasoning that there had to be something wrong somewhere. In that same year, 1935, this doubt led him to the famous thought experiment with the cat. Initially intended as a bad joke to ridicule the absurdities of quantum mechanics – especially superposition and the notion that observation influences outcomes – the experiment gradually became a running joke among quantum sceptics. Unsurprisingly, the idea for this experiment had actually been inspired by Einstein . . .

As an aside, Schrödinger's views on locality were not quite so uncompromising. Unlike his friend Albert, Schrödinger did dare to accept that aspect, assuming a certain interaction that possibly/maybe/perhaps might yet exist between two particles. Schrödinger targeted his arrows not so much at locality as at the interplay between quantum and classical. Between micro and macro. There had to be something in there, something he could use to make fools out of the quantum theorists. And sure enough, there was: superposition.

In his famous experiment, Schrödinger places a cat in a sealed box containing a Geiger counter, a handful of radioactive uranium atoms and a flask of poison. It is possible that radioactivity will be emitted during the experiment, but it is equally possible that it will not. To be more precise, at any given moment the uranium atom is in a superposition of being in its original and its decayed state, the two possible 'branches' of the wave function. In the branch of the wave function where the atom decays, radioactivity is emitted, a hammer falls onto the flask, which shatters, releasing a poison, instantly killing the cat. In the other branch of the superposition, the uranium atom has not decayed and the cat keeps purring quietly. So if you want to know whether the cat is dead or alive, there's nothing for it but to open the box. Looking at the cat (the measurement) breaks the superposition, and randomness will decide on which branch becomes reality.

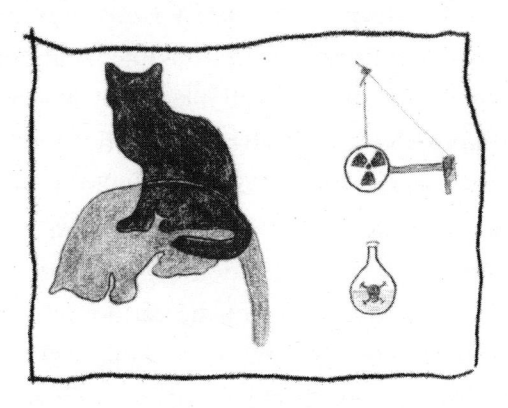

Schrödinger's cat

As long as the box remains sealed, the cat snuggles cosily in a superposition of dead and alive at the same time. If you open the box and look at its contents, you break the superposition and the cat is revealed to be either dead or alive. And although the probability that the uranium has decayed (and the cat is dead) rises as time passes, the outcome remains unpredictable and thus wholly dependent on chance. Such is the bizarre logic of quantum logic: a cat can be dead and alive at the same time.

Of course, Schrödinger took his experiment to the extreme by applying quantum theory right down to the letter. Except it's very simplistic to claim that everything can be both one thing and the other, as long as we don't look: here and gone, awake and asleep, president and forklift driver, tortoise and hare. But luckily we're almost at the halfway point of this book and the reader can now tell the difference between the hard facts and a spot of tom(cat)-foolery. Obviously, nothing can be both alive and dead. And how presumptuous it would be of us, mere mortals, to believe that the cat's fate depends solely on the moment *we* decide to open the box.

The paradox – because in this case we can genuinely call it a paradox – is that Schrödinger is overlooking an important detail. Just because *we* don't look doesn't mean no measurements are happening. It's impossible to isolate a box perfectly from its surroundings; there are always external agents, such as photons or molecules, that *can* interact with the box. These external elements, with their tiny microscope eyes, are constantly doing measurements. The point is this: in practice, the properties of the cat are fixed, whether or not someone opens the box to look with human eyes.

A brief look ahead. In the final few chapters we will see that the inevitable interaction between a quantum system and its environment is one of the biggest challenges in the development of the quantum computer. Quantum computers must perform many operations at once, relying on superpositions of very different states. But such superpositions can only persist when the system is completely isolated from its surroundings – a feat that is anything but simple.

And so, one answer inevitably gives rise to more unanswered questions. Questions to which we still have no conclusive answers. Where does superposition stop? At what point does something cease to be both/and, and turn into either/or? Where does the strange world of quantum physics give way to classical physics? Where does that boundary lie? And how exactly do you define it? These are questions that scientists from the second quantum revolution will explore in more depth. But the ultimate answer may remain forever out of reach.

The cat had barely recovered when Schrödinger added a coda to his take on the EPR article. In a later paper still in 1935, he became the first to introduce the concept of entanglement. We have already used this term liberally in this chapter, and for good reason. It is the only way to fully understand the experiments we've just outlined. The crux of the confusion between Einstein and Bohr was that neither of them knew entanglement existed, much less that it would become a cornerstone of quantum physics. Schrödinger, on the other hand, had grasped the phenomenon better than anyone else. And so, he was uniquely positioned to define it succinctly: 'the best possible knowledge of a whole does not necessarily include the best possible knowledge of its parts . . . I would not call that *one* but rather *the* characteristic trait of quantum mechanics, the one that enforces its entire departure from classical lines of thought . . . By the interaction the two representatives [quantum states] have become entangled.'

In summary: the whole is more than the sum of its parts. In an entangled system, knowing the state of each individual particle still tells us nothing about the state of the system as a whole. The critical information lies in the relationship *between* the particles, and not in the particles themselves. The truth lies, literally, somewhere in the middle. This is the main difference between quantum and classical physics.

5.6 The cat rings the Bell

Bohr had weakened Einstein's assumption, Schrödinger had strengthened it slightly, but it was Irish physicist John Bell who, thirty years later, would provide the clearest and most decisive understanding of how entanglement really worked.

The year was 1964. In the corridors of physics institutes around the world, the buzz was mainly about the practical applications of quantum physics in chemistry, nuclear science, particle physics and materials science. Hardly anyone was still concerned with the basic principles of quantum physics, or the interpretation of its increasingly peculiar laws. People had other cats to whip. But they hadn't reckoned with John Bell (1928–1990). During his free time at CERN,[1] Bell enjoyed tackling the unresolved conceptual tangles at the heart of quantum physics: 'I am a quantum engineer, but on Sundays I have principles.' For many of his colleagues, Bell was an unknown figure – until 1964, when they could no longer ignore him. With the publication of two surreally precise papers, Bell scattered fresh compost around the tender roots of quantum physics. Soon afterwards, the tree began to bear fruit.

In a first paper, Bell debunked a famous theorem by none other than von Neumann. *The* great von Neumann? The very same. Even the brightest minds sometimes miss the mark. According to von Neumann, no classical model existed that could reproduce the predictions from quantum physics. Bell, however, wasn't content to take this claim at face value. Maybe (who knows!) Einstein was a teeny bit right after all? Bell set to work and constructed a classical

---

[1] Conseil Européen pour la Recherche Nucléaire, the world's leading experimental centre for research into elementary particles, based in Geneva.

'hidden-variable model'. Surprisingly enough, the predictions of this model were indistinguishable from those made by the quantum physics of a qubit. Conclusion: von Neumann's claim was wrong.

To be clear, 'hidden variables' refers to the information that is unknown prior to the moment of measurement. This information dictates how particles behave when measured, as if they somehow 'know' in advance how they should respond. It's almost as if they've tacitly agreed among themselves how to act. This is the essence of what Einstein is referring to when he proposed the existence of an unseen, underlying theory – a theory that must help to explain the inexplicable, spooky behaviour of particles.

In his second paper, Bell showed how von Neumann's theorem could still be saved. For this could indeed be done, provided the classical theory was supplemented by an additional component: locality (the notion of separate realities, Einstein's hobbyhorse). Starting from Einstein's local realism, Bell constructed a mathematical model and gave it a new name: '**local** hidden-variable theory'. Bell's genius lay in realizing that a classical 'local hidden-variables model' could never be rich enough to reproduce the correlations observed in entangled quantum states. To corroborate his suspicion, he devised an experiment that would allow him to measure these correlations – and definitively reveal whether quantum entanglement could be reconciled with classical local realism.

HIDDEN VARIABLES THROUGH A DIFFERENT LENS

When your tax notice drops through your letterbox, you don't immediately know whether you'll get a refund or owe money. You don't find that out until you tear open the envelope and read its contents. At the same time, the content of the letter (the 'property') has been fixed ever since the calculations were completed and it was printed at the tax office. That content cannot and will not change, whether you open the envelope right away or leave it sealed for a week.

Things become more intriguing when there are two envelopes, as this helps to illustrate what happens in a local hidden-variables model. In the case of two classical envelopes, one person will receive a refund while the other will owe money. Here too, the notice is fixed in advance. It is not until the envelopes are opened that it becomes clear who the lucky one is, Alice or Bob.[1] And, crucially, if Alice gets money back, Bob will inevitably have to pay extra. Let's now look at quantum envelopes. When Alice opens hers, the content isn't already fixed in the classical sense. Depending on how Alice views her letter – say with or without rose-tinted spectacles – she might receive a petition about a controversial wind farm, a tax notice or news about her pregnancy test. The message she reads will be entirely random: either positive or negative. All we know is that the content of Alice's letter will always be the opposite of the content in Bob's letter – at least, if Bob performs the same type of measurement on his envelope.

Bell's arguments are not easy to explain, as they use the language of mathematics in a subtle and creative way. But let's give it a try. Bell constructed a model, using a local hidden-variables theory, simulating a system of two classical spins that are far apart. For each of

---

[1] Alice and Bob are fictional characters and a recurring duo in quantum sketches.

these spins you can independently perform one of two types of measurement: either in base 'a' or in base 'b'. Each measurement can produce two outcomes: 0 or 1. Thus, a single experiment results in one of sixteen possible scenarios (each spin can be measured in two different bases, each of which in turn can have two different outcomes: $2^2 \cdot 2^2 = 16$). For example, in one experiment, particle 1 is measured in base 'a' and the outcome is 0. Then particle 2 is measured in base 'b' and the outcome is 1. There are sixteen variations in total. A crucial aspect of Bell's theory is that the choice of the measurement (whether base 'a' or base 'b') must be entirely random and should not be agreed or communicated in advance. Moreover, the measurements should be performed independently by two different observers. The intention is to repeat the experiment over and over, each time starting from a randomly chosen measurement base.

John Bell

Once the experiments are complete, a table is created recording how often each of the sixteen scenarios occurred. By adding together eight of these numbers and subtracting the other eight, we obtain a value x. 'Bell's inequality' dictates that, in the case of a local hidden-variables model, the value of x must remain within a certain limit. The catch is that Bell proposes a quantum experiment involving two well-defined entangled particles (qubits in a 'Bell state') and two different measurement set-ups (two for Alice and two for Bob). This setup also results in sixteen possible scenarios, because each measurement on a qubit has two possible outcomes. If you total the outcomes in the same way as in the Bell inequality – summing eight of the outcomes and subtracting the other eight – you find a value greater than the limit set by the Bell inequality. In short: if the result of your experiment is a value greater than x, this proves that there is no local hidden-variables model that describes this experiment (and thus, nature itself). The good news is that this experiment definitively shows that quantum physics is not simply a cut-and-pasted 'local' classical theory embellished with a few additional hidden variables. The essence is that, if a hidden-variables theory is local, it cannot possibly be compatible with quantum physics. If it does agree with quantum physics, it must be non-local.

With Bell's thought experiment, Einstein's dream was forever smashed to smithereens: there is no all-encompassing classical theory capable of making the same predictions as quantum physics. Only an actual experiment could show whether the local-realist worldview to which Einstein was so attached would have to cede to the reality of entanglements. But that was still fifty years in the future. When it came, the result eliminated all doubt: the correlations predicted by quantum physics were indeed experimentally observed. This happened in 2022, when John Clauser, Alain Aspect and Anton Zeilinger could finally pop open the long-chilled champagne. That year, the trio were awarded the Nobel Prize in Physics for their relentless efforts since the 1970s to demonstrate the importance of John Bell's visionary publications. Their groundbreaking experiments pushed the boundaries of what was possible and paved the way for the second quantum revolution. They were able to confirm experimentally that quantum physics cannot be explained by a (classical) local hidden-variables model. With no exaggeration, John Bell can rightly be hailed as the great hero and founding father of the second quantum revolution.

Among the most science-fiction-like outcomes of Bell's work is quantum teleportation, another thought experiment, but this time devised by the author's Cambridge office neighbour Richard Jozsa and his collaborators. The concept of quantum teleportation is directly tied to the principles of entanglement and an extension of Bell's experiments. Teleportation is, again, a consequence of the fact that a measurement on one entangled particle has an instantaneous effect on its partner, no matter how far apart they are. Action at a distance doesn't get any spookier. If it hadn't already happened, reading about teleportation theory in his *New York Times* would have left Einstein forever spinning in his grave.

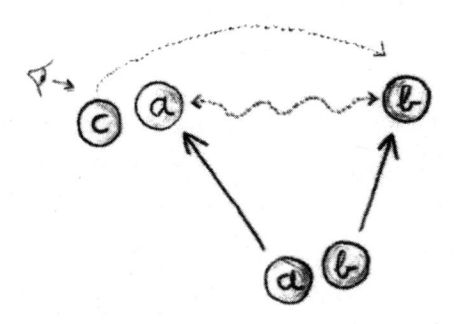

Quantum teleportation: particle c willingly allows itself to be transported by particle a to particle b.

Suppose particle a and particle b are entangled, and particle c is currently hanging out with particle a but would rather be teleported to particle b because the view is nicer over there. By conducting an entangled measurement – a joint observation involving both the particle to be teleported (c) and one of the entangled particles (a) – it is possible to teleport the quantum state of particle c to the location of particle b. This could be in another lab halfway across the world and would happen instantaneously.[1] You only need to get two particles to interact, so that this entanglement can then be used as a 'quantum bridge' allowing other particles to interact and transfer states later on. This might sound very futuristic, but quantum teleportation is one of the key concepts in the field of quantum communication and quantum computing.

Experiments with quantum teleportation of qubits have now become commonplace and have been successfully demonstrated in many a lab around the world. Next in line for a trip in the teleportation machine are whole atoms. As for when your corgi can take a spin in the teleporter, you'll just have to paws and wait.

5.7 Contextuality

> *Things in respect to themselves have, peradventure, their weight, measures, and conditions; but when we once take them into us, the soul forms them as she pleases. [. . .] Health, conscience, authority, knowledge, riches, beauty, and their contraries, all strip themselves at their entering into us, and receive a new robe, and of another fashion, from the soul; and of what colour, brown, bright, green, dark, and of what*

---

[1] The instantaneous nature of teleportation requires a footnote. This has to be qualified, of course, as information cannot travel faster than the speed of light. The teleportation step is not complete until the recipient has also received two classical bits for each qubit being teleported. Based on that classical information, the recipient knows in which base the received qubit is encoded, and how the qubits are to be interpreted.

quality, sharp, sweet, deep, or superficial, as best pleases each of them, for they are not agreed upon any common standard of forms, rules, or proceedings; every one is a queen in her own dominions.

— MONTAIGNE, *THE ESSAYS*, ON DEMOCRITUS AND HERACLITUS

Grete Hermann (1901–1984) was an original Noether girl. After studying mathematics and completing her PhD under Emmy Noether in Göttingen, Hermann shifted her focus to philosophy, where she enthusiastically built upon the foundations of physics. Long before John Bell, she had recognized that something was wrong with von Neumann's theory of hidden variables. Her insight stemmed from the inherent unpredictability of measurement outcomes for quantum particles. This led her to develop the concept of 'quantum contextuality': the idea that the properties of a quantum system depend entirely on the way we look at it and are shaped by the questions we ask (the 'observables'). Hidden-variables models, on the other hand, are by definition non-contextual: the properties of a system are fixed, independent of the nature of the measurement. Hermann had come to this realization in 1935, but her insight seemed to go unnoticed at the time.

It wasn't until thirty years later that Simon Kochen (b. 1934) and Ernst Specker (1920–2011) formalized quantum contextuality in a watertight mathematical formula.

Kochen and Specker refuted Einstein's realism according to which a system has properties that are an intrinsic part of that system, independent of our observation. They had made a very specific (and mathematically very complicated) construction allowing different types of measurements to be performed on a system. According to quantum physics, some of these measurements could be performed simultaneously (commuting observables), while others could not. In the hidden-variables model, the outcome of all these different measurements has to be fixed in advance. However, Kochen and Specker demonstrated that this 'counterfactual reasoning' (cast your mind back to the Bohr–Einstein debate) leads to a conflict with the predictions of quantum physics. Specifically, they showed that in quantum physics, certain measurement results will show opposite values depending on which commuting observables are measured. Conclusion: the outcome of measurements depends inherently on the type of measurement (the 'context'). We can assign a value to measurements only when we also know their context.

From a philosophical perspective, the contextuality of quantum physics led to a huge rupture with classical physics. Observing a system is no longer a detached act; we do not observe as outsiders. A measurement result is influenced by all other measurements and thus has no objective, independent value; it depends entirely on how you measure or look at it. Everything we see is the result of the impact that our observation inevitably has on a measurement. In quantum physics, the observer is by no means above, or outside, the experiment: the 'I', with its human eyes, is an integral part of it. This is entirely in line with the Heisenberg microscope. Incidentally, Hermann and Heisenberg had a regular coffee date.

Philosopher Hannah Arendt was deeply impressed by these quantum insights, incorporating them into her broader philosophical framework. She remarked: 'The answers of science will always remain replies to questions asked by men; the confusion in the issue of "objectivity" [the idea that man would stand outside the experiment and observe everything from outside without influencing it] was to assume [incorrectly] that there could be answers without questions and results independent of a question-asking being. Physics, we know today, is no less a man-centred inquiry into what is than historical research.'[1]

---

[1] Hannah Arendt, *Between Past and Future*, Penguin Books, 2006, p.49.

IN A NUTSHELL

> My milkshake contains no better atoms than yours. All elementary particles are exactly alike and they are either bosons or fermions.

> The wave function of many particles contains boundless degrees of freedom. Physicists perfect the art of approximation. Our scouts are Hartree, Fock and Feynman.

> Trophies of quantum: the electronic structure of atoms, the formation of chemical bonds, the band structure that gives materials their properties, no transistors without quantum.

> Bose party with Einstein.

> *Dramatis personae*: Satyendra Nath Bose, Albert Einstein, Douglas Hartree, Vladimir Fock, Richard Feynman, Enrico Fermi, Wolfgang Pauli, Linus Pauling, John Bardeen.

SIX

ONE, TWO, MANY

Let's recap. In Chapters 1 and 2, we outlined the mathematical basis on which quantum physics is built. Chapter 3 covered the history of how Planck, de Broglie, Bohr and Einstein turned quantum physics into a transformative science that gradually made itself indispensable for the description of the smallest particles that make up nature and their interactions. Particles are waves, waves are particles; it all seemed too crazy for words, but it is what it is. In Chapter 4, we explored the mathematical formalism needed to explain all these curiosities and came face-to-face with the enigmatic phenomenon of superposition. Chapter 5 showed us that, as soon as we examine a system consisting of multiple particles, even stranger things happen. Nature, it seems, isn't afraid to step on the toes of our intuition. The bizarre behaviour of entangled systems made even Einstein's hair stand on end. Yet, what was demonstrated experimentally had to be unmistakably right. Truths fell by the wayside and even the sharpest minds occasionally found themselves humbled. Nevertheless, Stevin's theory has endured – not just through centuries but through chapters of this very book.

This chapter is also about entanglements, albeit in a more

constructive way. Instead of staring blindly at the supposed 'mysteries' of nature, we now harness quantum physics to uncover how matter is fundamentally structured. To do so, we must appeal to the physics of many particles. That is great news, because this opens the door to a whole host of new discoveries: how atoms turn into molecules during chemical reactions, and (in Chapter 7) how nuclear physics pushes the boundaries of smallness even further.

Inevitably, this will raise countless new questions. Will the sun ever burn out? Where does it get its heat from? What exactly are superconductors and transistors, those linchpins of all virtual life on Earth? Why is green green? And, oh yes, while we're at it: how old is the Earth? These are questions that scientists have spent years pondering. And the answer is invariably quantum.

David Mermin summed it up rather nicely: 'Quantum mechanics works. Indeed, no theory of physics has ever had such spectacular success. From ignorance about the structure of matter, quantum mechanics has brought us, in less than a century, to an understanding so broad, powerful, and precise that virtually all contemporary technology relies on it. And the theory has enabled us to make sense of phenomena far beyond anything technology has yet been able to exploit.'[1] And this, despite the fact that we've only scratched the surface of quantum physics.

6.1 The indistinguishability of particles

Until now, we have performed experiments on one or two particles, done measurements on one or two particles and provided the

---

1 N. D. Mermin, 'Making better sense of quantum mechanics', *Reports on Progress in Physics*, 82, 012002, 2018.

description of one or two particles. This approach enabled us to understand how a system works as a whole. But that broader picture, called matter, consists not of a few particles, but of a humongous number of particles. This shift introduces two entirely new concepts that are fundamentally different, but equally important. First: the wave functions of many-particle systems have exponential complexity. These systems are described in an exponentially large Hilbert space – a mathematical framework that serves as the collective 'home' for all possible wave functions – requiring an exponential number of variables to specify them.

WAVE FUNCTIONS OF MANY PARTICLES

One particle is described by a wave function $\psi(x_1)$, whereby x_1 represents the coordinate of the particle. Each point in space corresponds to a complex number. Two particles have a wave function with two variables $\psi(x_1, x_2)$. For each possible position of the first *and* the second particle together, the wave function gives us another complex number. Three particles have a wave function with three variables, such as $\psi(x_1, x_2, x_3)$, and each possible combination of x_1, x_2 and x_3 yields another complex number. And so on. This pattern continues, and the complexity grows rapidly. Even if each particle is limited to just two possible positions (such as a qubit), the wave function of *n* qubits must still be specified for an exponential (2^n) number of different configurations, namely for all possible positions of the *n* particles.

ONE, TWO, MANY

The second factor to consider when it comes to describing many-particle systems is the fact that identical particles are indistinguishable. In the quantum kitchen, you can't say: this is asparagus 1 and this is asparagus 2. Asparagus 1 can equally be asparagus 2. However, these veggies can be in a superposition of both asparagus and parsnip. Of course, this defies any rational explanation, making it well worth taking a closer look. Measurement results depend on the wave function. Consider two indistinguishable particles. If particle 1 is in position 1 and particle 2 in position 2, there's no meaningful distinction if we swap them. Particle 1 could just as well be in position 2 and particle 2 in position 1. Since the particles are identical, the wave function remains permutation invariant, meaning unchanged. Physicists express this a bit more eloquently, saying that the wave functions 'transform in a trivial way under the permutation group'. For the record, a permutation group (Galois's hobbyhorse) is the set of all possible ways in which a number of objects can change order. The permutation symmetry of a wave function can be achieved in two ways. In one case, the wave function is symmetric: if we reverse (any) two particles, the wave function remains unchanged. Mathematically, this looks like $\psi(x_1, x_2, x_3) = \psi(x_2, x_1, x_3)$. This symmetry applies to bosons (with the photon being the most well-known example). Phenomena such as lasers and Bose–Einstein condensates owe their existence to this symmetry. More on this later.

In the other case, the wave function is antisymmetric (or more eloquently: it 'transforms in a projective way'): if we reverse the two particles, the wave function is preceded by a minus sign (e.g.: $\psi(x_1, x_2, x_3) = -\psi(x_2, x_1, x_3)$). This behaviour is a hallmark of fermions (with the electron, the proton and the neutron being the most well-known examples). Even if a wave function is multiplied by a minus sign, the predictions will be exactly the same as those with the original wave function. The reason is that probabilities are always calculated by squaring the wave function, which eliminates the sign.

There is a kind of redundancy, a superfluousness, in the description of the wave function. Of course, there is only one number that equals 'minus itself', and that is zero. Four is not the same as minus four. But zero is the same as minus zero. If you have a function that is antisymmetric in two variables (for example, the position of one particle relative to the position of another particle, or the orbital of one particle and the orbital of another particle), when the variables equal each other, the function must be zero. In other words, the probability of two particles being in the same energy state (or: two electrons in the same orbital), is simply non-existent. Zero. The mystery of why fermions repel each other to such an extent, and are therefore responsible for the stability and hardness of matter, can be explained by antisymmetry. Antisymmetry – or the exclusion principle – is basically the strongest force in nature. *Pauli!*

The existence of virtually everything discussed in the following pages is directly or indirectly indebted to this exclusion principle.

THE NAME OF THE GAME

Fermions and bosons share a common story: both are named after the mastermind who first took them seriously. Fermions owe their name to Enrico Fermi (1901–1954). Together with (but independently of) Dirac, Fermi derived a formula, known as Fermi–Dirac statistics, that can be used to determine the energy levels a system of fermions occupies as a function of temperature. Bosons are named after the Indian mathematician and physicist Satyendra Nath Bose (1894–1974), who was the first to describe the behaviour of this other indistinguishable particle, using Bose–Einstein statistics.

6.2 Hotel Hilbert

The more particles, the more numbers (or 'variables') you need to describe the wave function. And that is exactly where the problem lies. The number of variables does not simply increase, it increases exponentially. With every additional element, the total number of variables doubles. That makes it, in the end, virtually impossible to describe a material. By way of comparison: the number of particles in a material is roughly equal to Avogadro's number, named after Lorenzo Romano Amedeo Carlo Avogadro, Count of – take a breath – Quaregna and Cerreto (1776–1856). This unit determines how many molecules twelve grams of carbon contain. And this is $6.02214076 \cdot 10^{23}$, as many as the number of grains of sand in the Sahara Desert (give or take). Now imagine a system of 10^{24} qubits. Describing such a system would require $2^{10^{24}}$ variables. It quickly becomes impossible to even conceptualize numbers of that magnitude.

ATOMOS IS INDIVISIBLE

One of the most baffling episodes in physics is the fact that by the end of the nineteenth century, some physicists already knew there had to be such a thing as atoms, and they even knew how big these atoms were, how much they weighed, what their average velocity was in a gas at a certain temperature, and how many atoms one gram of hydrogen contained – all this without anyone ever having directly observed an atom.

The protagonists in this episode are James Prescott Joule (1818–1889), known for the famous (or despised) calorie information on our food, Johann Josef Loschmidt (1821–1895), the underappreciated father of modern chemistry, and the ubiquitous James Clerk Maxwell (1831–1879). Starting with Ludwig Boltzmann's hypothesis that matter consists of indivisible atoms (*atomos* is the Greek word for 'indivisible'), these pioneers developed a microscopic theory that explained the laws of thermodynamics.

Their reasoning unfolded as a kind of relay race: Joule deduced the velocity of gas particles by relating pressure in a gas to the change in the momentum of the particles when they collide with a wall (this velocity is independent of the individual mass of the particles). Maxwell related the 'mean free path' – the distance particles travel before colliding with other particles – to the velocity at which gases mix (this velocity was determined experimentally). Loschmidt had the crucial insight that the diameter of an atom could be related to the ratio of the volume of a liquid to that of a gas, multiplied by this 'mean free path'. Finally, Maxwell set everything out in a magnificent article in *Nature*, in which the true size and weight of atoms were determined with remarkable precision.[1]

Despite the overwhelming success of these atomic theories, they faced fierce opposition from the scientific establishment. The Austrian physicist and philosopher Ernst Mach seized every opportunity to ridicule Ludwig Boltzmann, the great propagator of atomic theory. Boltzmann, who could no longer take the constant attacks, was left despairing and purposeless. He took his own life in Duino (in present-day Italy) in September

---

[1] 'Molecules', *Nature* 8, 437–41, 1873. Abstract: 'An atom is a body which cannot be cut in two. A molecule is the smallest possible portion of a particular substance. No one has ever seen or handled a single molecule. Molecular science, therefore, is one of those branches of study which deal with things invisible and imperceptible by our senses, and which cannot be subjected to direct experiment.'

1906. And that's regrettable on two counts. Because thanks to Einstein's second paper (dating from 1905 and relating to Brownian motion), all doubts about the existence of atoms were finally and definitively dispelled.

Bear in mind that a wave function of 500 qubits (which actually describes a relatively small number of particles) has no fewer than 2^{500} variables. That's more than the total number of atoms in the universe. Who could possibly do calculations with numbers like that? The sheer complexity of describing a system of many quantum particles is both maddening and profoundly challenging. Personally, we find it infinitely fascinating that something so agonizingly incomprehensible can produce so many comprehensible phenomena. Anyway, we are not done with infinity yet. It's all a lot more complex than that.

EXPONENTIALS OF EXPONENTIALS

If you want to calculate how large the Hilbert space is or how many distinguishable states exist in the qubit world, beware: this number does not increase exponentially, but doubly exponentially. That's right, by the exponential of an exponential. By comparison, imagine an epidemic that increases doubly exponentially. Instead of each person infecting around five people, they would infect as many people as the total number of people already infected at that time. It would spread awfully quickly.

Let's return to our 500 qubits. If we want to calculate the number of distinguishable states in which these qubits can be

> found, we're faced with the mind-boggling calculation $x^{2^{500}}$, whereby x is a parameter representing the precision of our measurements (the more precise, the bigger x is). To visualize this, imagine a very large sphere and plenty of very small spheres. Now calculate how many of those very small spheres fit into the very large one. Oh yes, that very large sphere has not three, but 2^{500} dimensions. The number of very small spheres that fit into it is therefore not just exponentially large; it's ludicrously large.

The problem with many particles is that every time a particle is added, the dimension of the Hilbert space doubles. However, this so-called Hilbert empire is more like a castle on a cloud, an illusion. We can juggle with exponentials of exponentials, but the number of states that nature can create by itself is – and always will be – limited. And that's the consequence of one of the fundamental properties of nature: all interactions are local.[1] Despite the seemingly infinite expanse of that Hilbert space and all calculations we can theoretically perform within it, the number of states that nature can produce on its own *is* comprehensible. Nature, and everything needed to do physics, is just an exponentially small back room of that exponentially large Hilbert space. Only that little corner corresponds to the physical world.

---

1 That is to say: interactions in nature always occur between two particles. As a result, it takes a very long time before complex states arise in nature. By comparison: if you want some news to reach as many people as possible as quickly as possible, you can do so via, say, social media. You can reach millions of people in just one click or post. In the quantum world, however, everything happens via word of mouth. One person can only inform one other person at a time. So, of course it will take much longer for the news to spread.

Everything else is not physical. And it is exactly that small back room that scientists want to (try to) understand. If not today, then tomorrow – just maybe.

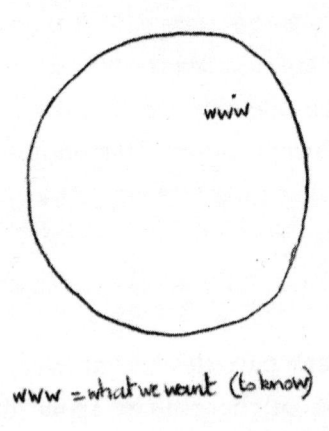

www

www = what we want (to know)

Hilbert space, a hotel with an infinite number of rooms in which all possible wave functions reside – with a tiny antechamber as the only space accessible to us (and to nature).

Conclusion: we need a different way of doing physics, because it is simply impossible to describe many particles with such a jumble of variables. Dirac may not have been one for many words, but in 1929 he hit the nail on the head with his statement: 'The fundamental laws necessary for the mathematical treatment of a large part of physics and the whole of chemistry are thus completely known, and the difficulty lies only in the fact that application of these laws leads to equations that are too complex to be solved.'

Since its inception, quantum physics has been a succession of attempts to approximate what exactly that hidden back room of nature looks like, and what's going on in there. We will outline two methods that were developed to better understand that back room: the Hartree–Fock method and Feynman diagrams. Thanks to these approaches, it will be possible to understand a lot about chemistry, as well as solids, nuclear physics and quantum field theory. If you have ever browsed a popular science book on quantum physics before, you may have noticed that these methods are conspicuously absent from the table of contents. Are they too technical? Or too difficult? They are certainly very technical, and we can by no means call them easy, but let's be honest, everything here is technical and unreasonably complicated. What makes them worth mentioning is that they make amazingly good sense. They reveal fundamental truths about the structure of matter, making them indispensable in quantum physics. In fact, we cannot do many-particle physics without them. The question of why one method works for a given problem and another doesn't, is as important as the question whether a particle is a boson or a fermion. And although they are incredibly efficient, scientists continue to search for even better methods, because there are still enormous gaps and grey areas in the numerous systems where these approaches *don't* work. These so-called strongly 'correlated systems' represent the core challenge in modern research on many-particle physics.

THE HARTREE–FOCK METHOD

We owe much of our knowledge of modern physics to the Hartree–Fock method (named after its inventors, Douglas Hartree [1897–1958] and Vladimir Fock [1898–1974]), discovered shortly after that key year of 1925. This method enables us to express wave functions of many electrons in an extremely efficient way. The core idea is to approximate a many-particle wave function as a product of single-particle wave functions, ensuring that the overall wave function remains asymmetric. What makes this approach so effective is that it takes into account the most important (and most powerful!) force at play: Pauli's exclusion principle. But there is more. The forces at work between electrons (via the electromagnetic force) are also partially considered – albeit in an approximate manner – helping to tame some of the exponential complexity.

To understand what the wave function of electrons around an atomic nucleus looks like, the Hartree–Fock method dictates that the many-electron wave function is a product of single-particle S, P, D and/or F orbitals. This simplification works remarkably well. Exceptionally well, even – and luckily for us, for without it we would never have been able to understand the physics of atoms.

The Hartree–Fock method provides a way to approximate the wave function and energy of a system composed of many fermions. Thanks to this method, we can look deep into the soul of atoms, crystals, semiconductors and metals – topics we will explore in greater detail later in this chapter. However, there's a limitation. With this method, the full wave function is a sum of the individual wave functions, meaning the electrons are not entangled with each other. Yet in nature entanglements *do* tend to be important. The Hartree–Fock method falls short when you want to make quantitative predictions in terms of very precise numbers. And since quantum physics is mainly concerned with predictions, we need something else. Time for Richard Feynman.[1]

RICHARD FEYNMAN AND HIS DIAGRAMS

Before we introduce Richard Feynman (1918–1988) and his diagrams, we need to make an essential stop-off at the ground state. First of

---

[1] Feynman with a single 'n' at the end, please. His colleague Gerard 't Hooft shows his admiration for Feynman in a rather interesting way. If he receives e-mails in which Feynman's name is spelled with a double 'n' at the end, he immediately consigns them to oblivion.

all: what we deem a comfortable room temperature is very cold for electrons (or fermions). On the other hand, bosons find room temperature far too hot. This disparity implies that fermions and bosons require completely different physics.

For now, let's focus on electrons/fermions. When electrons are 'very cold', they collectively settle into the lowest possible energy state (ground state), huddled and snug near the atomic nucleus. However, in a system with many particles, the structure of their collective wave functions gets wildly complicated. The ground state is less a peaceful resting place and more like a bubbling, chaotic soup: a lively mess of entanglements, ripples and particles that appear and disappear at will, like oil bubbles in a lava lamp. This absolute nothingness (a.k.a. the vacuum) is so staggeringly complex that understanding it is one of physics' greatest challenges. If this is clear, everything else is as easy as π.

The question remains: how can the ground state of a system of interacting particles be described by perturbing the solution of the Hartree–Fock method?

In practice, as Newton predicted, forces are always the result of the exchange of particles. The fact that two equal charges repel each other, for example, is due to the photons they exchange. In other words, fermions pass on their entire accounts in terms of mass, charge and energy to each other via other intermediate particles.

To make this more tangible, Feynman invented a rather intuitive technique, known as Feynman diagrams. These diagrams look almost childishly simple, but that was precisely Feynman's talent: presenting and explaining highly complex ideas in a simplified way. The problem is that many people confuse these diagrams with reality, whereas they are simply an abstraction. For Pauli, they were nothing other than pathetic little drawings. Now, that smart alec Pauli could say what he liked, but the diagrams had the intended effect, in the sense that they made it possible to do the lengthy, confusing and seemingly impossible calculations needed to describe the interactions between particles. Feynman diagrams provide an insightful glimpse into the complexity of the ground state, that notorious back room of Hilbert space.

Let's zoom in on a simple diagram illustrating the electromagnetic force between two electrons (see below). The x-axis shows the motion in space. The y-axis shows the motion in time. The straight lines trace the trajectories of the fermions (two electrons, in this case) and the wavy line represents the path of a photon.

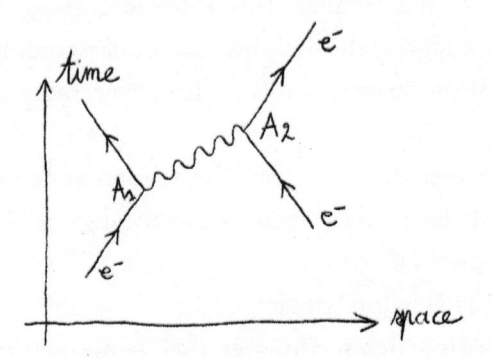

Feynman diagrams depict the forces at work between two electrons.

In this diagram, an electron (e⁻) emits a photon at point A1. This photon is reabsorbed a little later (at A2) by another electron. Upon emission of the photon, the first electron is propelled backwards in space, while the second electron, upon absorption the photon's energy and momentum, is deflected in the opposite direction. This interaction causes the particles to repel each other. Returning to our explanation of the ground state, we know that anything can happen between points A1 and A2. Virtual particles can disappear into it or they can even be pushed back in time. The number of possible interactions between particles is infinite, shaping the so-called 'vacuum soup'.

The following example illustrates how complex that soup can taste. At point A1, a photon (p) is converted into a fermion-antifermion pair (electron and positron); these are annihilated at point A2, turning back into a photon that is then ultimately absorbed. The lower version shows that the fermion-antifermion pair, *en route*, exchange yet another photon at point x:

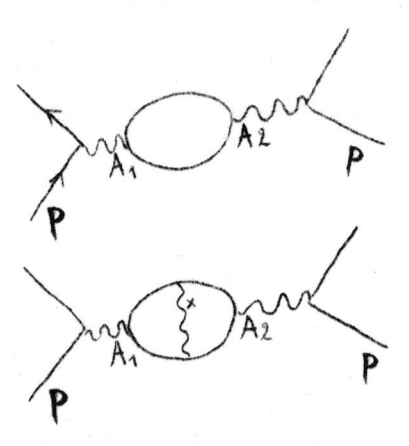

In their simple ingenuity, Feynman diagrams reveal one of the most astonishing qualities of quantum physics: the ability to teach us things about particles that have not yet been discovered, particles we can't even see, but which are present virtually in the diagrams. Their predictive power is extraordinary. The diagrams predicted, for example, the existence of a new elementary particle, the charm quark, and even estimated its mass.[1] It was only by assuming the existence of this charm quark – where it functioned as a virtual particle – that the theory aligned with experimental

---

1 Sheldon L. Glashow, Jean Iliopoulos and Luciano Maiani, 'Weak interactions with lepton-hadron symmetry', *Physical Review* D 2: 1285, 1970.

results. These diagrams provide us with a remarkably clear view of the bubbling, stewing vacuum within a system of interacting particles: it is a ceaseless cycle of creation and annihilation of particles and antiparticles.

Feynman was well aware of the limitations of his perturbative technique, particularly when interactions became too strong. He began dreaming of a method that could directly describe the correlations between all particles without the need to map out the entire wave function.

To develop such a method, a new language is required, where qubits and entangled Bell pairs form the vocabulary, and where the grammar is defined by entanglement entropy and tensor networks. This is fodder for the twenty-first century, and therefore for Chapter 9.

Now that we have understood the principles of many-particle physics, we can apply the theory to relevant problems.

Richard Feynman had a personalized Dodge Tradesman Maxivan. Because QED and QUARK had already been taken, and because a personalized number plate could contain a maximum of six letters, he chose: QANTUM. The boot featured a diagram of two muon neutrinos exchanging a particle. At the time, Feynman had no idea what that particle could be. Many road trips later, this particle was proven to exist, and named the Z boson.

6.3 Atoms and molecules

Time for chemistry. Chemistry examines how atoms bond to form molecules and materials. During chemical reactions, the unique, hidden (or secret) properties of matter literally reveal themselves. Toss wood onto a fire, and the conversion of cellulose into CO_2

releases heat. Mix carbon dioxide with water, nitrogen and oxygen, and you'll end up with a bar of dynamite in your hands. Iron with oxygen gives rust, sugar over a fire becomes caramel, leaves use the sun's energy to convert CO_2 and water into sugar and oxygen, and an alchemist could even convert lead into gold. Incidentally, Newton wrote more about his 'first love' than about physics. He compiled a sprawling index of thousands of pages filled with texts, authors, references and keywords related to the magical and mysterious science called alchemy. But he never managed to distil a coherent theory from it. That's not too surprising, since chemistry is, essentially, applied quantum physics.

A LOT WITH A LITTLE

If we look around us, we see spoons made from wood, cars made from metal, buildings made from glass, barracks made from stone, skirts made from fabric, little boats made from paper. We also see countless living organisms: plants, bacteria, algae, animals, and so on. Things that all appear completely different. And yet, their composition can be reduced to a specific set of atoms; all matter in the world consists of a remarkably limited number of chemical elements. These are the elements that Mendeleev arranged in his periodic table according to their weight and shared properties. Each element has an atomic number corresponding to the number of protons in its nucleus, which in turn corresponds to the number of electrons orbiting that nucleus. Beyond number 83, the nuclei become increasingly unstable due to nuclear forces, and beyond 92 (uranium), all bets are off. They can be made in a lab, but they won't live long.

MENDELEEV'S TABLE

Dmitri Mendeleev (1834–1907) must have taken the saying 'Actions speak louder than words' to heart when he set out to construct his periodic table in 1869. Frustrated by his inability to clearly explain his ideas to his chemistry students at St Petersburg University, he sought a more visual and systematic approach. His goal was simple yet ambitious: to arrange all known chemical elements based on their properties.

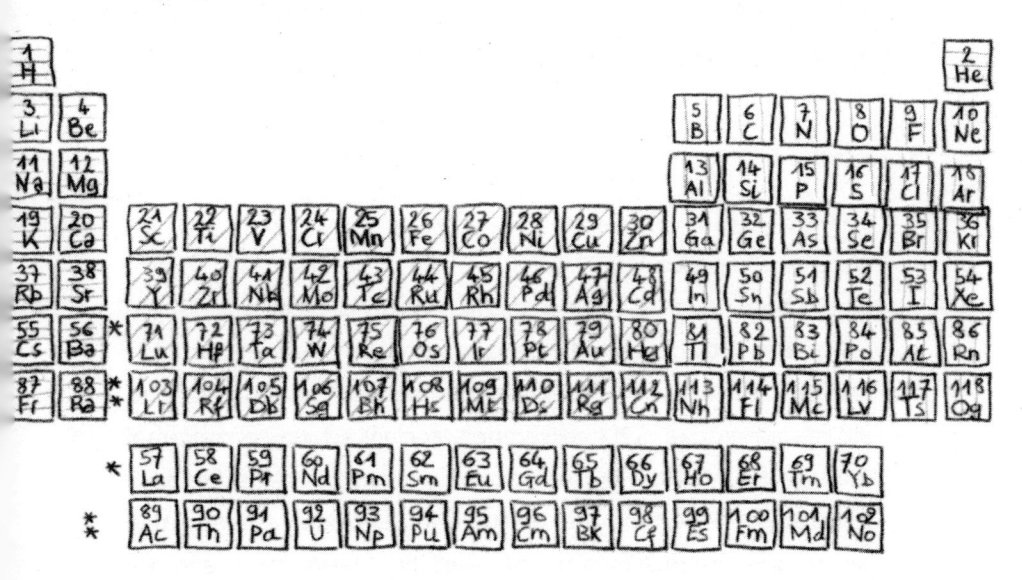

The periodic table

It must be said: the initial version of the table contained quite a few blank spaces. These gaps indicated elements that had not yet been discovered. But Mendeleev soon revealed himself as a master of heuristics. With uncanny accuracy, he predicted the existence of as-yet-unknown elements, such as gallium and germanium, which were later proven to exist. For the rest, he knew there had to be many other beautiful things in store, but as for what . . . he couldn't (yet) say.

Mendeleev eventually grew old and died, leaving others to gradually fill in the gaps in his table. Whereas about sixty-three elements were known during his time, the count now stands at 118 – and there are undoubtedly many more yet to be discovered. It was later revealed that all atoms in turn consist of just three even smaller particles: electrons, protons and neutrons. Atomic physics is like a Matryoshka doll: each doll contains an even smaller one. And eventually, we arrive at the tiniest, indivisible doll: the quark.

SIZES AND WEIGHTS

Bottles of Russian Standard vodka proudly declare that 'Dmitri Mendeleev, the greatest scientist in all Russia, was commissioned by the Tsar to establish the Imperial quality standard for Russian vodka'. Of course, the standard 40 per cent alcohol content for vodka was determined much earlier – in 1843, to be precise – when the pint-sized young Dmitri was barely nine years old. The Russians of the early twentieth century had a flair for embellishing national pride, liberally mixing fact with a splash of exaggeration and clever marketing. But claiming that Mendeleev standardized vodka is so wildly bamboozling, you'd half expect someone to follow up with a tale saying your grandfather is too old to have fought in the Hundred Years' War. What is true, however, is that Mendeleev's doctoral dissertation was titled 'On the combination of alcohol and water'. That said, vodka didn't warrant even a drop of ink in

> his thesis. It is also true that later in his career, Mendeleev was appointed director of the Russian Central Bureau of Weights and Measures. In that capacity, he was responsible for introducing the metric system in Russia.

The first major breakthrough in quantum physics came when Mendeleev's periodic table could finally be explained through a mathematical model. Until the discovery of quantum physics, chemistry was a cauldron of uncertainties. But the fog lifted when Pauli used the Hartree–Fock method to decipher the structure of the table. Suddenly, it became possible to calculate the energy levels of all atoms, understand the bonds between them and predict chemical reactions.

THE ATOMIC MODEL

In order to determine the possible energy orbitals of one negatively charged electron around a positively charged nucleus, the Schrödinger equation must first be solved for one particle. The key to solving this puzzle is, as you might expect, symmetry. In essence, the problem is equivalent to finding all the fundamental tones (S, P, D and F) of a three-dimensional drum, as discussed in detail in Chapter 2.4; the same mathematical framework, but with a radically different interpretation.

According to Pauli's 'trick' (incorporating the electron's spin), an electron can be in each energy orbital in two distinct ways: with a spin-up and with a spin-down. Consequently, S orbitals get a two-fold degeneracy, P orbitals a six-fold (three times two), D orbitals a ten-fold and F orbitals a fourteen-fold one.

The S, P, D and F orbitals describe how the wave function

transforms under rotations. However, the Schrödinger equation also provides another value: the 'radial properties' of the wave function, which describe how the wave function changes as a function of its distance from the nucleus. This radial component of the wave function is represented by a natural number. The larger this number, the greater the probability of the electron being further from the nucleus.

Each possible energy orbital is therefore represented by four symbols: a natural number, S, P, D or F, a value indicating which degenerate orbital the electron occupies, and its spin. Arranged according to increasing energy, this produces the following orbitals (not taking into account the spin and degeneracy, since the energy does not depend on these): 1S, 2S, 2P, 3S, 3P, 4S, 3D, 4P, 5S, 4D, 5P, 6S, 4F, 5D, 6P, 7S, 5F, 6D, 7P.[1]

With the exception of the hydrogen atom, all atoms have more than one electron. The exact number can be found in Mendeleev's periodic table. But what does the wave function of such a multi-electron system look like? That description is, generally, exponentially difficult. The wave function can be approximated using the Hartree–Fock method, which renders this wave function a product of single-particle wave functions. Compared to a single-electron system, these orbitals are much more tightly clustered around the nucleus, because the nuclear charge of these atoms is significantly higher than that of the hydrogen atom.[2] The Hartree–Fock method takes

---

[1] If you look carefully at the list, you see that the 4S precedes the 3D orbital. To be fair, this is cheating, as this is not the case for the single electron case. But when you turn on interactions, 3D electrons acquire more energy than the 4S electrons. This is why 4S precedes 3D in the periodic table.

[2] This also explains why virtually all atoms are approximately the same size: the greater the positive charge of the nucleus, the closer the first orbitals are to the nucleus. The same applies to the number of orbitals. The more orbitals there are around the nucleus, the more closely packed together they are. Fortunately, too,

the interactions between electrons into account, albeit to a very limited extent: the electron orbitals influence one another, but they do not entangle. The closer we get to the boxes at the bottom right of the table, the less well the Hartree–Fock approach works, as the orbitals here contain many more electrons, and interactions play a much more important role.

To make sense of the wave functions of different atoms in the periodic table, we'll start at the bottom – filling the lowest-energy orbitals first and working our way up until every electron has found its place. Once the first S orbital (1S) is filled (and that happens quite quickly, because it only has room for two electrons), a second S orbital (2S) with slightly more energy is filled. Next comes a P orbital (2P) followed by another S (3S).[1] And so on.

A quick guide to reading the table: the number in each box represents the number of electrons surrounding the atomic nucleus, and is equal to the number of protons in the atomic nucleus. This number determines the atom's weight. Elements in horizontally striped boxes (in the table on page 185) have an S orbital as their outermost electron shell. For the elements in boxes with vertical stripes, the last filled orbital is a P orbital. In the boxes with diagonal stripes, this is a D orbital, while unshaded

---

because otherwise it wouldn't take long before atoms would assume gigantic proportions.

1 Electrons have other eccentric behavioural characteristics too. There can be two electrons in each of the three orbitals of a P orbital. You might think that electrons will therefore occupy P orbital after P orbital, but nothing could be further from the truth. Electrons first separately occupy one orbital each. The first goes to the first P orbital, the second electron goes to the second P orbital and the third electron goes to the third P orbital. Initially, therefore, all of them are alone. If a fourth electron arrives, it joins the first P orbital, a fifth electron joins the second P orbital and a sixth goes to the third P orbital. Sorted. When electrons occupy one orbital each, they also all have the same spin. If a second electron is added, it has the opposite spin. That's the way it is. This is known as Hund's Rule.

boxes correspond to F orbitals. The two bottom rows technically belong in rows six and seven (indicated by the asterisks * and **), but they are placed below to prevent the table from becoming too wide.

A few examples:

→ H (hydrogen) has one electron in the 1S orbital.

→ He (helium) has two electrons in the 1S orbital.

→ Li (lithium) has two electrons in the 1S orbital + one electron in the 2S orbital.

→ N (nitrogen) has two electrons in the 1S orbital + two electrons in the 2S orbital + one electron in each of the three 2P orbitals.

→ Ti (titanium) has eight electrons in the four first S orbitals + twelve electrons in the P orbitals + two electrons in the D orbitals.

The most relevant electrons are on the outer shell, determining the atom's chemical properties and how it bonds with other atoms. Situated at the greatest distance from the nucleus, they are also the most free to move. Elements stacked in the same column of the table exhibit remarkably similar properties because their outer shells share similar numbers of electrons and types of orbitals (S, P, D or F). Such is the ingenious design of the table.

In its box, each atom comes with a set of extra numbers. One of these numbers refers to the binding energy (or ionization energy). This is the energy required to remove an electron from an atom (i.e. to ionize it). Ultimately, this is what happens in a chemical reaction: electrons are plucked from one atom so that they can be planted in an orbital of another atom (or multiple atoms). Nothing is destroyed or created in the process; everything is merely restructured.

Atoms within the same vertical column typically share a similar ionization energy: they exhibit a 'periodic' regularity. Atoms with fully filled outer shells have particularly high ionization energies – they cling tightly to their electrons and are highly resistant to giving them up. Noble gases (found on the far right in the table), for instance, are famously non-reactive because their electrons are too tightly bound to their nucleus. Removing them would require an enormous amount of energy. As a result, a noble gas will always remain a noble gas.

Another number in the table relates to electronegativity. This is, essentially, the opposite of ionization energy. It is the energy you gain by adding an electron to the atom. Both ionization energy and electronegativity can be determined experimentally, and the results align surprisingly well with the predictions of the Hartree–Fock method.

When everything in the table finally clicked into place, one thing was unmistakably clear: quantum physics is *the* theory. Why? Because the predictions of quantum physics were not only qualitatively correct (elements with similar properties belong in the same column), but also quantitatively (allowing exact calculations of ionization energy, electronegativity, polarization and magnetic momentum). Today, this table, complete with its structure, numbers and interpretations, is the ultimate reference; a veritable bible for every chemist.

DISHING THE DIRT (2/2)

Pauli had a somewhat troubled relationship with chemistry, though he likely had only himself to blame. In 1929, he took the plunge and got married, but the union sank within a year. His bride had run off with another man. Pauli reportedly said that if it had been with a bullfighter, he might have been able to stomach it. But a chemist?!

FILLING ORBITALS: MOLECULES

In much the same way that electrons organize themselves to form atoms, we can analyse how atoms arrange themselves to form molecules. Until the discovery of quantum physics, the mechanism behind atomic bonding remained a complete mystery. 'There must be a new force at work!' Wrong. By 1925, it became clear that no new force was needed. All chemical bonds could be explained by incorporating the electromagnetic force into the quantum mechanical description of the atom.

Chemical reactions occur when the outer energy orbitals of different atoms fuse together to form one molecule. The sum of all the outer energy orbitals is called the outer 'shell'. But what exactly happens during the formation of a molecule? When atoms combine, the electrons in their outer shells no longer orbit just their own atomic nuclei, they also orbit the nuclei of the other atom(s). In this way, the total energy drops. In other words, by bringing the atoms close together, they gain energy. However, if they get too close, the innermost electron orbitals repel each other very strongly (Pauli again!). As a result, atoms tend to settle at an 'ideal distance' from each other. This determines the size of a molecule. The stability of a material or molecule is directly proportional to the energy saved.

The greater the energy reduction, the stronger the bond. That's precisely what drives nature to form matter: it always seeks to minimize energy. Very democratic, Mother Nature!

The study of noble gases' stability demonstrated that fully filled shells are especially important from an energy perspective, which leads to the following rule of thumb – and actually the only rule you need to know to understand the basics of chemistry: *atoms form bonds with other atoms so that the electrons they share lead to completely filled shells.*

By way of illustration: two hydrogen atoms (each with one electron) can fuse to form one hydrogen molecule. This results in one large S orbital twirling around the two nuclei together, which is fully occupied by two electrons, each with opposing spin:

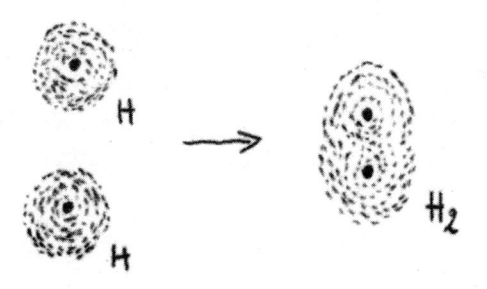

Two hydrogen atoms fuse to form a hydrogen molecule.

Since the two atoms are identical (H), their outer orbitals have the same energy. These orbitals move alongside/underneath/over the top of each other and form a superposition of left-plus-right or left-minus-right states. In both cases, the electrons in these orbitals are completely delocalized between the two atoms – once again, they are everywhere and nowhere at the same time. In the minus orbital, the electrons are more tightly packed, which increases their kinetic energy (thanks, Heisenberg!), creating an energy difference between the plus and minus orbitals.

The graph below shows how the energy levels of these orbitals evolve as a function of the distance between the two atoms.

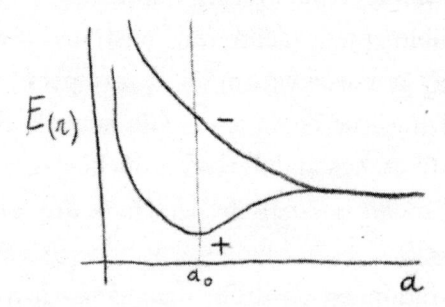

x-axis: the distance between the atoms; y-axis: the energy as a function of that distance for the plus and minus orbital; a_o = the distance with the least energy, or: the distance between the two atoms in the ground state

Things get even more interesting when we look at nitrogen (N), whose outer P shell contains three electrons (one in each of the three P orbitals). To completely fill its outer shell, the nitrogen atom needs three additional electrons. It can 'borrow' these from three hydrogen atoms. But this borrowing is reciprocal, as the hydrogen atoms each 'borrow' an electron from the nitrogen to completely fill their own S shell. That's two birds with one stone, because both the P shell of nitrogen and the S shells of the hydrogens end up completely filled. The result is ammonia (NH_3), a very stable molecule.

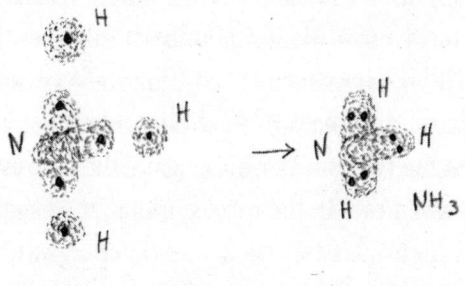

Ammonia, NH_3

Oxygen (O) has room for only two additional electrons in its outer P shell, allowing it to bond with two hydrogen atoms. Together, this forms H_2O (water), the most abundant molecule on Earth. Here, as in NH_3, both the S orbitals and the P orbitals are fully occupied. However, in this case, nature takes an intriguing detour towards a state of lower symmetry. Since the filled 2S orbital of oxygen has roughly the same amount of energy as the outer 2P orbital, they merge into a single SP orbital (a process known as hybridization) with four arms and eight compartments for storing electrons. These arms form a perfect tetrahedron. In the case of oxygen, two of these orbitals are completely filled, and the other two have one electron each. The latter will therefore in turn bond with the 1S orbitals of hydrogen.

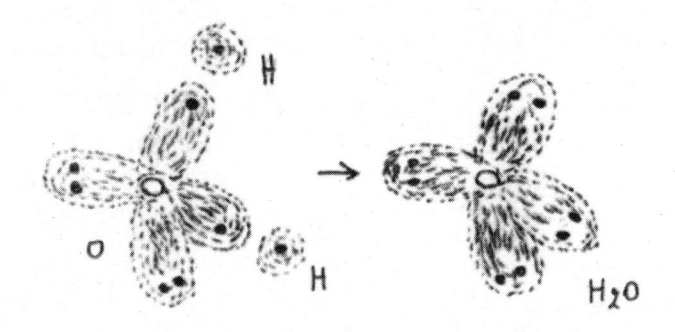

H_2O: hybridization of the S and P orbitals

Carbon (C) is in a league of its own, as it has only two electrons in its outer P orbitals, leaving room for four additional electrons. Keeping the rule of thumb in mind, carbon can share two P electrons with an oxygen atom, resulting in carbon monoxide (CO), which, unfortunately, is responsible for many an incident involving malfunctioning stoves. Furthermore, carbon, like oxygen, has a filled 2S orbital with roughly the same amount of energy as the outer P orbital. Linus Pauling (1901–1994), the godfather of quantum chemistry, was the first to recognize that this hybridization leads to an extraordinary variety of complex bonds. This versatility is why carbon forms the backbone of all life on Earth. By definition, the organic branch of chemistry – its largest branch – focuses on molecules containing carbon atoms. Through SP hybridization, carbon doesn't just share two electrons but four, bonding with two oxygen atoms, which results in carbon dioxide (CO_2).

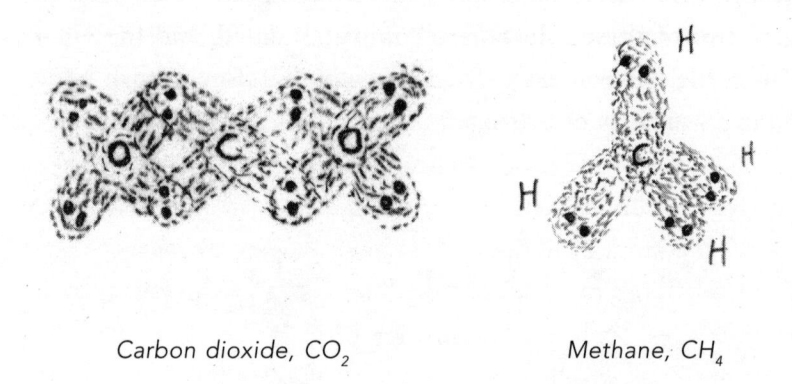

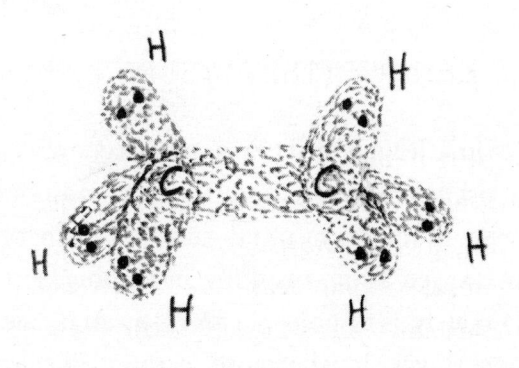

Carbon dioxide, CO_2

Methane, CH_4

Ethane, C_2H_6 or $H_3C - CH_3$

A little imagination is all it takes to see that CH_4 (methane) is also a remarkably stable molecule. Similarly, two carbon atoms can bond with six hydrogen atoms (as in ethane: $H_3C - CH_3$), or they can form a chain with n carbon atoms and 2 . (n+1) hydrogen atoms (propane, where n = 3; butane, where n = 4; octane, with n = 8, etc.). These are the so-called hydrocarbons.

Smaller molecules of this type are gases, larger ones are liquids and very long ones form solids (like candle wax), all of which are ideally suited for use as fuel.

We can also modify the carbon-hydrogen bond by adding an oxygen atom, resulting in molecules such as alcohol: C_2H_5OH. Or we can replace two hydrogen atoms with one oxygen atom. The result: fatty acids, glucose (sugar) and cellulose (the basic structural component of wood). Amino acids, such as glycine and alanine, which form the basis of all living matter, are obtained by replacing one hydrogen atom at the end of a chain with an NH_2 group. The ends of these types of molecules can then in turn easily bond with other molecules, leading to the formation of proteins. And so the process continues, producing ever larger, longer and more complex molecular structures.

AUGUST KEKULÉ (1829–1896)

In a not so distant past, the German chemist August Kekulé (to be pronounced with an accent on the last 'e' at his father's insistence) was the leading chemist of his time. Kekulé had found a method to graphically represent chemical bonds between molecules – a remarkable feat considering that no one had ever seen a molecule or atom back then. A bit like Feynman's 'pathetic little drawings'. With elegant simplicity,

Kekulé depicted the atomic composition of a substance, how the atoms were bonded and how many electron pairs were shared. Kekulé is best remembered for discovering the atomic structure of benzene (C_6H_6), a colourless organic compound with a faintly sweet aroma. Benzene is a primary component of petrol. While researchers already had a solid empirical understanding of benzene's properties, its precise structure was still a complete mystery – how exactly the six carbon and six hydrogen atoms were arranged, and how the electrons navigated through this setup. The solution came to Kekulé in a daydream about a snake biting its own tail (the Ouroboros). *Aber natürlich!* Benzene molecules are arranged in a ring structure! What's unique about this bond is that the electrons are in a superposition of alternating single and double bonds. This allows them to swing freely from one electron orbital to the next, like monkeys swinging from vine to vine.

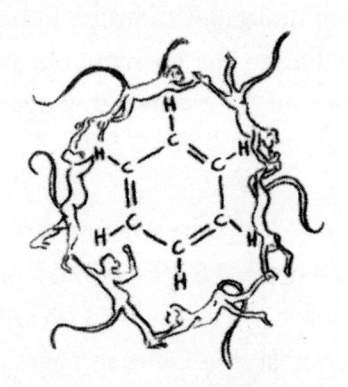

Benzene, represented by August Kekulé's graphical method. Dashes represent electrons from the outer shells, responsible for the chemical bonding.

Let's turn our attention to chemical reactions – more precisely, combustion. What happens when something burns? Take methane

(CH_4) as an example. In combustion, methane reacts with two oxygen molecules (2 . O_2) to produce carbon dioxide (CO_2) and two water molecules (2 . H_2O). This can be represented as: $CH_4 + 2 . O_2 \rightarrow CO_2 + 2 . H_2O$. The end product has a lower energy than the starting product. This energy difference is released as heat, visible and tangible to us as fire (a form of electromagnetic radiation). This reaction unfolds in two distinct phases. First, energy is needed to split CH_4 and O_2 into nine individual atoms: $CH_4 + 2 . O_2 \rightarrow C + 4 . H + 4 . O$. As this requires quite a lot of energy, this reaction does not happen spontaneously. On the other hand, once initiated, the generated heat helps break apart additional molecules more easily. Next comes the rearrangement: the individual atoms form new molecules: $C + 4 . H + 4 . O \rightarrow CO_2 + 2 . H_2O$.

What about wood? Trees have leaves. The chemical process of photosynthesis uses energy from sunlight to convert CO_2 and water into cellulose (wood = $C_6H_{10}O_5$). In doing so, a tree quite literally plucks its mass out of the air. Wood thus contains more energy than CO_2 and H_2O, making it an excellent reservoir of solar energy. When that same tree trunk ends up in a fire, the wood is converted again into CO_2, releasing its stored energy. However, the cellulose must first be broken down into individual atoms – hence, lighting a fire isn't always straightforward. Once the wood catches alight, the chemical reaction sustains itself until all the wood is consumed. Conclusion: the warmth of your cosy fire is pure solar energy – and nothing is lost.

The combustion of food in our stomach follows the same quantum mechanical process, albeit at much lower temperatures. During digestion, very long molecules are broken down into smaller pieces. While these longer molecules yield less energy, the barrier to initiating the chemical reaction is much lower, allowing it to occur at not-so-high temperatures. This is because the relevant frequencies in long molecules are much lower than in short ones, so the energy required to initiate a chemical reaction is much smaller (remember Planck's formula $E = h \cdot v$).

Why, then, is there life on Earth and not on Mars? The answer lies in Earth's ideal conditions for life to develop in all its complexity. It is just cold enough here for chemical bonds to remain stable, but just warm enough for chemical reactions, such as photosynthesis, to occur spontaneously. If it were a few degrees warmer, molecular vibrations would tear chemical bonds apart, making such reactions impossible. If it were a few degrees colder, everything would freeze into a vast, lifeless stasis of infinite stillness. Another vital factor is the Earth's atmosphere, which blocks almost all (lethal) cosmic rays. No other planet boasts an atmosphere so perfectly suited to this purpose. Who, then, would dare to contradict Leibniz when he reflected that we truly 'live in the best of all possible worlds'?

6.4 Hard matters

In a world filled with water or gas, chaos reigns. This chaos exists because the structure of the atoms or molecules looks exactly the same from every perspective. When it starts to freeze in that world of water or gas, order arises. Lev Landau described it far more poetically: symmetry breaks. Order, in essence, is nothing more than fractured perfection. It's a beautiful phrase, though seemingly contradictory. That's because here, we are using the mathematical interpretation of the word 'order'. When it freezes, the world no longer appears uniform from all directions on an atomic level. As chaos 'breaks', matter rearranges itself into one of the 230 possible configurations atoms and molecules can adopt. According to group theory, there are precisely 230 ways to break translational and rotational symmetry to form a crystal lattice – no more, no fewer. Perhaps most striking is that nature has found a way to produce a material for each of those 230 variations.[1] No wizard could conjure such a feat. In pre-quantum times, it was a significant mystery how – or why – a material chose one of the 230 possible forms to adopt. The advent of quantum physics brought an answer: the crystal structure is entirely determined by electrons in the outer shells of atoms. As atoms move closer together, their outer shells merge, and the electrons stop orbiting a single atomic nucleus. Instead, they spread across vast electron orbitals, sometimes encompassing an enormous number of atomic nuclei. This merging creates a continuum of possible energies, known as bands, which may or may not be separated by an opening, or 'gap'. In these bands, the electrons behave like a very peculiar type of gas. So why is matter hard? Because even in this gas (or: in these bands) only one particle can occupy each energy level. What's more, the electrons orbiting these fused orbitals create strong bonds that hold everything tightly together, making the material exceptionally robust and resistant to being pulled apart. Just like in atoms and molecules, the different energy levels are filled one by one, starting with the lowest. The illustration overleaf depicts how the energy levels of electrons in a crystal vary as a function of the distance between the atoms. The further you move along the a-axis, the greater the distance between the atoms.

---

1 There are 230 possibilities for breaking symmetry in our three-dimensional world. In two dimensions, there are seventeen ways to break symmetry, and therefore seventeen possible tessellations: the 'wallpaper groups'. In one dimension, there are exactly seven ways, the 'frieze groups'. In four dimensions, there are 4,894.

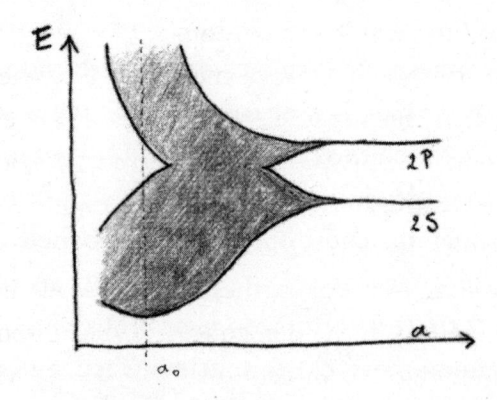

a = the distance between two nearby atoms in a lattice; a_o = the distance with the lowest total energy (the distance in the ground state)

In the above diagram – reading from right to left – we see an S orbital and a P orbital fuse into two 'bands', separated by a 'gap'. These two bands fuse, and then split again. Each of these bands is a collection of compressed orbitals that are filled one by one with electrons (starting from the lowest energy level). The precise shape of this band structure depends significantly on the type of crystal. The distance between atoms in a material corresponds to the distance (a) at which the total energy of all electrons together is at its lowest (a_o). So which crystal structure do the atoms eventually choose? The one where the total energy of all the electrons is minimized. But with 230 possibilities, talk about choice overload!

Depending on the number of electrons in the outer shell of the atoms, a band is either fully occupied or not. If the band is completely filled, we call the material an insulator. If it isn't, it's a conductor. For conductors, very little energy is required to nudge an electron into a new orbital – they are a bit more accommodating in such matters, as there is no gap to overcome. When an electric field is applied, the electrons reorganize themselves into orbitals that mainly have a momentum pointing in the same direction as the field. And because there is room in the band, they can do so without too much effort. Since most electrons then 'go with the flow' rather than against it, electrical current is created. Common conductive materials include silver, gold, copper, aluminium, mercury, steel and iron.

For insulators, however, the situation is entirely different. Here, the bands are fully occupied (these are the valence bands), with a 'gap' beside them (a 'forbidden zone', where electrons cannot exist) followed by an empty conduction band. Since moving an electron from the valence band to the conduction band requires energy equal to the total energy of the 'gap', a very strong electric field is necessary for the electrons to reorganize. That's why materials with this property, such as diamond, are utterly dull – they simply don't conduct.

In chemistry, as the rule of thumb made clear earlier, fully filled bands are very advantageous from an energy point of view. The same principle applies to the band structure. Think of it like a bicycle: when your tyres are pumped up to their maximum capacity, leaving no room for even a breath of extra air, you save a lot of energy while pedalling. This analogy helps explain why insulators occur in nature. Nature favours a crystal structure in which all bands are neatly filled. That's easier said than done, because it's not always possible to find a crystal where the total number of energy levels in the bands matches exactly the total number of free electrons in the atoms. If, despite the numerous options, the band isn't completely filled, the material becomes a conductor.

If we take a closer look at the band structure, we gain a clearer picture of what is happening there. Crystals have a translational symmetry (see lesson 1 of Emmy Noether's goldfish: shift a crystal, and its structure still looks the same). We label the closely packed electron orbitals forming a band (or continuum) by assigning each a new number (k) that represents how these orbitals transform under translations. This number is known as the crystal momentum, a variant of the momentum.

The illustration below provides a graphical representation of the band structure of silicon. On the x-axis we find the crystal momentum; on the y-axis, the possible energies for electrons with this value of k. The Greek letters on the x-axis refer to specific values of k, which have a particular symmetry under reflections. The lines below zero (the dotted line on the y-axis) represent the energies of the orbitals in the valence band; the lines above zero represent the conduction band. Sandwiched between them is a very thin 'gap'. This illustration of the band structure of silicon, derived using the Hartree–Fock method, heralded the start of the digital revolution.

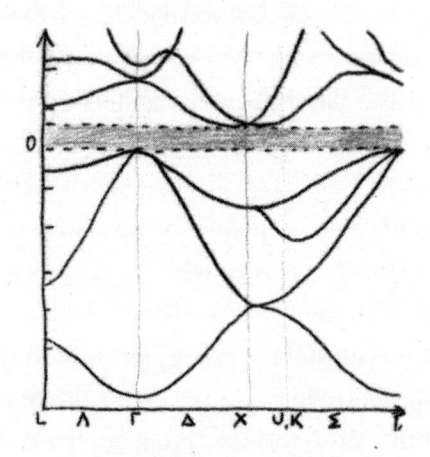

The band structure of silicon; the thick solid line is the 'gap'.

The most notable feature of this band structure is its relatively narrow 'gap'. Quantum physicists John Bardeen (1908–1991) and Walter Brattain (1902–1987), along with engineer William Shockley (1910–1989), all working at the renowned Bell Laboratories, saw immense potential in this. By introducing impurities into the silicon crystal, they discovered that it could conduct better than an insulator, though still not as efficiently as a conductor. These impurities introduced a small number of energy levels within the gap, providing electrons with a springboard to hop from the valence band to the conduction band. The result was a semiconductor. This invention, in 1947, led to the most disruptive technological breakthrough of the twentieth century: the transistor, a portmanteau of 'transfer' and 'varistor' (variable resistor). This discovery would have been completely impossible without a thorough understanding of the quantum physics of electrons.

Initially, transistors replaced the very fragile and energy-guzzling radio lamps (also known as vacuum tube amplifiers) of the early twentieth century. Today, they are indispensable in technology. They are everywhere. Nearly every human on Earth carries billions of transistors with them every day. A single smartphone, from a certain well-known fruit-branded company, contains no fewer than 15 billion (!) transistors.

DOPING WORKS

Semiconductors are insulators, the difference being that the former have small impurities in the 'gap' between the filled band and the conduction band. These impurities are actually 'forbidden' energy orbitals that arise when a material is 'doped'. Depending on the nature of the doping, these artificial electron orbitals are either filled (and then we get an n-type semiconductor) or left unfilled (p-type). In a transistor, an n-type is paired with a p-type, creating a kind of microscopic switch that determines whether the current can flow or not.

> The result is either a one or a zero. On or off. These miniature switches are the fundamental building blocks of every application of a computer.

A vital spin-off of semiconductors is the solar panel, which turns sunlight into electric current. Sunlight provides electrons with such a jolt of energy that they leap merrily from the valence band to the conduction band. The energy gap between these two bands determines the voltage of the solar panel, powering your smartphone, computer, or even your twinkling Christmas tree, illuminated by yet another quantum phenomenon: LED lights. LEDs, or 'light emitting diodes', are a simplified cousin of the transistor. They are far more sustainable and energy-efficient than incandescent bulbs, converting most of the energy into light, rather than frittering it away as heat. LEDs illuminate every corner of modern life, from the subtle glimmers on your devices to calculator screens, luminous billboards, television sets and even the unnecessarily dazzling video backdrops accompanying your favourite artists as they sing and swing on stage. Green and red LEDs led the way for years, but it was the long-awaited blue LED that truly completed the spectrum, allowing white light to shine at last.

6.5 Quantum colour

This brings us to another phenomenon that is ever-present in our daily lives yet impossible to explain without quantum physics: colour. We may have everything our hearts desire, but why does the grass always seem so much greener on the other side? A better question, perhaps, is why grass is green in the first place. What determines an object's colour? Objects have a colour because our eyes can perceive light of different wavelengths and filter out those beyond the light spectrum. The colour of light is determined by the frequency (or wavelength) of the electromagnetic radiation emitted by everything around us.

Electrons at the surface of a material absorb photons, causing them to jump to a higher energy level. They only jump to a higher energy orbital when they have absorbed a photon with exactly the right energy ($E = h \cdot v$), being the energy difference between the two orbitals. If a photon strikes the material with a frequency that would excite an electron into a forbidden zone (the 'gap'), this photon cannot be absorbed and is reflected. The colour is thus determined by the frequencies that *disappear*. The photons that are *not* absorbed by the material are reflected, giving a material its colour. Clear.

So why are plants green? Because photosynthesis is powered by the molecule chlorophyll. Chlorophyll absorbs blue and red light, converting it into energy. Green light, however, is not absorbed. Instead, it is reflected and flies straight into our eye. Another example: copper. Blue photons have enough energy to lift electrons in copper to a higher band, but red light particles do not. Hence copper's warm reddish glow. One more, for good reason: the vibrant hues that flowers use to woo insects. Some 70 per cent of the incoming light is absorbed by the flavonoids and carotenoids, two pigment types in the flower. They decide the fate of the light. The reflected light paints peonies pink and poppies red. The same alpha- and beta-carotene play a vital role in processes that determine whether our skin looks pale or tanned, and in the production of vitamin A (via photosynthesis). Truly, our bodies are huge quantum factories at work!

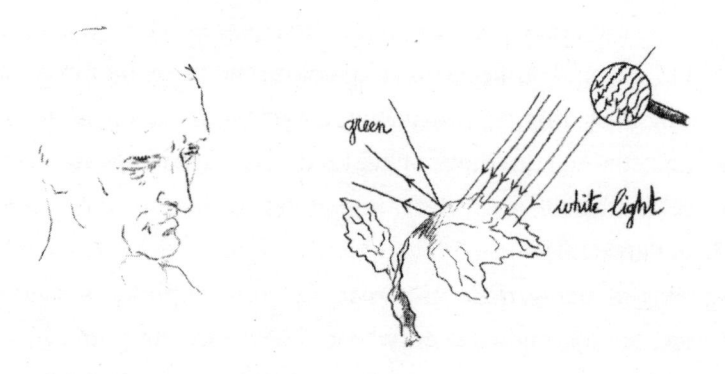

The colour of a material is determined by the light rays that are not absorbed, but bounced back.

NEWTON VERSUS GOETHE

Once upon a time, literary salons were all the rage. Poets, writers, philosophers, musicians, politicians, scientists and other curious minds would gather to sip coffee, share cake, argue passionately, crack jokes and emerge a little wiser. In those days – we're talking about the seventeenth and eighteenth centuries – every self-respecting intellectual dabbled a little in physics, just as they might in music or literature. In these salons, people showed their true colours.

Through the refraction of light by a prism, Newton demonstrated that white light was composed of all colours. And since this was Newton's view, who dared question it? Enter Johann Wolfgang von Goethe (1749–1832). Goethe, a fierce adept of the salons (where he discovered Newton's theories), did something Newton had never done in his entire life. He travelled. To Italy. And what he saw there, left him awestruck. No, it wasn't the girls. It was the colours. Those colours! Overwhelmed,

Goethe pushed aside his other obsession (Faust) and dedicated a significant part of the rest of his life to the study of light.

The main difference between Goethe and Newton was their focus. Newton sought the nature of light; Goethe, the essence of colour. For Newton, light was made of particles. Goethe saw it differently. He believed that humans – not lab instruments – were the most precise tools for studying nature. His phenomenological approach stood in stark contrast to Newton's reductionist approach. According to Goethe, colour arises from the interplay between light and darkness, with a hazy medium mediating between them, and from two fundamental hues: yellow and cyan. Light tinged by darkness becomes yellow, and darkness brightened by light turns blue. From this interplay emerges what Goethe called 'darkened light' and 'lightened darkness'. Goethe's theory brimmed with nuance. He distinguished warm colours from cool, paired each colour with its complementary hue, and linked these shades to human emotions. Light brought goodness and joy; darkness symbolized badness and sorrow.

Newton and Goethe: two figures, two worldviews. Newton gave us a theory; Goethe offered a sensibility. One relied on experiments; the other, on experience. Both were grounded in deep intuition, yet they diverged. For Goethe, it's not about nature, but about the way nature is experienced, the interpretation – and that's something very personal.

What Goethe did wasn't hard science, but that didn't make it irrelevant. As Heisenberg said: 'Goethe's colour theory has in many ways borne fruit in art, physiology and aesthetics.'[1]

---

[1] That wasn't the end of Heisenberg's reasoning. He continued: 'But victory, and hence influence on the research of the following century, has been Newton's.'

Even the choice of colour in shops, designed to tickle our consumer instincts, is inspired by Goethe's ideas! Life, after all, is more than exact sciences. Just because something isn't scientifically provable doesn't make it insignificant. Arguably, physics could be seen as the ultimate form of materialism: superficial and emotionless. It leaves no scope for nuance, poetry or emotions. But is that really true? Of course not! Physics too thrives on wonder, driven by a desire to uncover beauty, whether that beauty fits neatly inside the lines, or dances wildly outside of them.

6.6 On Bose, Einstein and lasers

Enough about fermions for now. Over to the universe of bosons. Unlike fermions, bosons are indistinguishable particles that can happily occupy the same state all at once. Most of them being weightless and inclined to gather en masse, they stand apart from the microscopic particles we've discussed so far: bosons manifest themselves in the 'normal' world. We *see* light. We *feel* the warmth of sunlight. The same goes for gravitons, which carry the force of gravity.

THE CORRESPONDENCE PRINCIPLE

In Newton's classical physics, particles were described as just that: particles. Not waves. Then quantum physics came along, and de Broglie revealed that particles are also waves. Light behaves as waves; that was the consensus. Until Einstein showed that they are also particles. But since photons are 100 per cent quantum mechanical in nature, how is it that James Clerk Maxwell could so accurately describe light and electromagnetism as early as 1873? How could radios be built back then, given that radio waves are made up of photons and are therefore quantum? How could experiments involving magnetism and electric fields be understood at all? Maxwell's electromagnetism is *the* theory that describes large numbers of light particles in terms of (classical) waves. That is to say . . . In order to understand this theory from the point of view of quantum physics, we need a humongous number of bosons. And when that many bosons are involved, the individuality of each particle disappears. Their quantization fades into the background, making all these particles resemble, quite simply, a classical wave again. It's like looking at a ladder in the distance. You can't make out the rungs any more; you only see one solid, flat, unbroken bar. From the point of view of quantum physics, the quantum states that occur in Maxwell's theory are special cases, called coherent states. Because the particles here do not interact with each other, they can be described using single-particle physics. This is in sharp contrast to interacting particles, which plunge us back into the complexities of Hilbert space and its unmanageable infinities. In this sense, the classical theory is a limit of the quantum theory of many independent particles, a tangible manifestation of Niels Bohr's correspondence principle. This correspondence principle implies that the quantum aspect of a quantum system with large quantum numbers is no longer really visible, allowing the system to be described using a corresponding classical theory. Here, the contrast between bosons and fermions becomes striking. Fermions, bound by the exclusion principle, cannot accumulate in large quantum numbers. Fermions

> always behave purely quantum mechanically; bosons do only sometimes.
>
> Despite their classical behaviour, bosons can also enter a state where their quantum mechanical properties become strikingly explicit, a state with no counterpart in classical physics. Once again, bosons love hanging out in the same state. If these bosons happen to be atoms cooled to extremely low temperatures, they can form a Bose–Einstein condensate.

THE BOSE–EINSTEIN CONDENSATE

Satyendra Nath Bose was the first to propose the idea of an indistinguishable and über-social particle in a paper. The paper in question initially failed to pass muster with the gatekeepers of academic publishing. Thankfully, after a circuitous journey, it found its way to Einstein's desk. The year was 1924. Einstein immediately realized that Bose had hit the proverbial nail on the head, even though the latter was blissfully unaware of the revolutionary theories unfolding in Europe at that time. Einstein arranged for a publication, though not before adding a few of his own annotations and insights.

Building on Planck's insights, Bose concluded that photons sharing the same energy level are indistinguishable. He then argued that their indistinguishability means they must be treated as one, and that they could consequently only count once in the statistics used to describe large collections of particles in statistical physics. That said, bosons, if the opportunity arises, will not readily turn down the invitation to shack up together. Out of all particles, they prove themselves the ultimate quantum partygoers.

ROLLING DICE WITH BOSONS

The probability of two particles being equal is higher in bosonic statistics than in classical, non-bosonic statistics. Let's illustrate this with a pair of dice. There are thirty-six possible outcomes when two dice are rolled. Of these, six outcomes are identical (1-1, 2-2, 3-3, 4-4, 5-5, 6-6), while thirty are different. According to classical statistics, the chance of the dice showing the same result is therefore 6/36 (or 1/6). According to bosonic statistics, the thirty throws in which the two dice show different values count as only fifteen (since 2-6 is considered equivalent to 6-2, meaning they count as a single throw); the probability of two dice being identical is 6/(6+15), or 2/7, almost twice as likely as in classical statistics. This inclination of bosonic particles to flock together explains why they are often found in shared states, unlike their classically governed counterparts.

On the basis of Bose's statistics, Einstein predicted the existence of a mysterious, undiscovered phase, something neither liquid, solid, nor gas; a phase where quantum particles become completely indistinguishable, blending seamlessly into one another without really interacting, to form what he described as a kind of quantum liquid.

Central in Einstein's mental leap was the assumption that the (bosonic) particles only interact very weakly. To put this into context: at very high temperatures, atoms always form a classical gas. If you reduce the temperature, that gas usually becomes a liquid. Lower that temperature further, and your system of strongly interacting particles will solidify into a hard substance. But what happens when you cool a system of particles with very weak interactions to the extreme? Einstein concluded that instead of crystallizing, these particles form a condensate: the Bose–Einstein condensate. This extraordinary phase, where particles fuse into a shared quantum state, could, according to Einstein, only be achieved under ultra-low temperatures. For experiments with rubidium atoms, for example, the temperature needs to plummet to a staggering 0.0000001°C above absolute zero. That is to say: -273.14999999°C. Even in the middle of cosmic nowhere, where temperatures hover at a relatively balmy 2.7°C above absolute zero, it feels toasty by comparison. Needless to say, this is not a place you can explore with a thick scarf, woolly socks and a flask of hot soup.

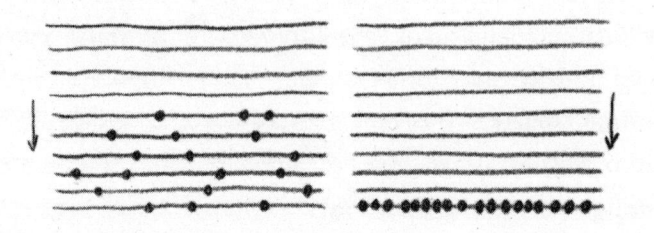

Particles condense, much like beads on an abacus or notes on a musical stave sliding down to rest on the bottom bar, merging into a single, stretched-out monotone.

'Why so cold?' you might be wondering. The answer lies in energy. As a system cools, particles naturally settle into their lowest energy state, because there's simply no room left in the higher levels. The phase transition occurs when a macroscopic number of bosons congregates in the ground state. At this point, bosons transform from discrete dots into overlapping waves. As their de Broglie wavelengths merge, they lose their individuality, huddle together in the same quantum state and entangle as if forming one fluid, unified whole. In this phase, particles undergo nothing short of an identity crisis.

If you cool a gas of rubidium atoms enough, they undergo an identity crisis and reach a state of enlightenment, merging into complete unity. Their individual distinctions fade, their minds clear and they settle into the serene nirvana of the absolute ground state.

Of course, creating such an extraordinarily low temperature doesn't happen with a simple flick of a switch. For this, experimentalists rely on a highly complicated setup involving state-of-the-art lasers, finely tuned magnetic fields, a healthy dose of patience, meticulous precision, and let's not forget, a hefty sum of money. Each of the steps and sub-steps of the cooling process must be executed flawlessly, down to the tiniest detail. And if you were thinking you could just copy and paste a Bose–Einstein condensate from someone else's work, think again. Not only must the instruments be perfectly calibrated, but even local variations in the Earth's magnetic field come into play. In short, cloning someone else's success is not an option.

It took an astonishing seventy years after Einstein's prediction for it to finally be realized in a laboratory. The first time was in 1995 in Boulder, Colorado (US), where scientists Carl Wieman (b. 1951) and Eric Cornell (b. 1961) achieved a Bose–Einstein condensate using rubidium atoms. Shortly thereafter, Wolfgang Ketterle (b. 1957) at the Massachusetts Institute of Technology (MIT) produced a Bose–Einstein condensate with sodium atoms.

In the seventy dizzying years that separated Einstein's theory from its laboratory confirmation, doubt and scepticism ran rampant. Many physicists refused to believe this could be true. Was the interaction between particles really so insignificant? Could particles genuinely interact so weakly? And yet, as history would show, Einstein was (once again) absolutely correct.

That said, much remains unclear. What exactly happens when matter transitions from one phase to another (the phase transition)? When does something stop being one thing and start being another? It's well established that during phase transitions, just like when a Bose–Einstein condensate is formed, fascinating and unusual phenomena occur. These events are deeply tied to symmetries – but not symmetries alone. They also involve 'critical exponents,' like emergence and universality, which we'll delve into later.

In the case of a Bose–Einstein condensate, a highly abstract form of symmetry breaking takes place: the phase of the condensate. In the Bose–Einstein phase, atoms stop behaving as independent entities and instead act as a single, unified superatom (or quantum field) perfectly synchronized in every way: the same speed, the same direction . . . What makes a system cooled to such extreme temperatures remarkable, is how vividly its quantum effects emerge. Since 1995, the Bose–Einstein condensate has become an indispensable tool in the development of quantum computers, atomic clocks and atom lasers, as well as in the exploration of a quantum theory of gravity and black holes. However, it remains, above all, a uniquely powerful way to simulate the behaviour of matter at the atomic level.

Even Schrödinger was initially lukewarm about Bose's work. It wasn't until he had lengthy discussions with Einstein and Planck that it struck him: Bose's paper wasn't simply an interpretation of Planck's theory, it went much further. The Bose–Einstein condensate strengthened Schrödinger's conviction that there had to be a new way to approach quantum physics. This realization planted the seed for his own brilliant equation, which he would derive one year later, in 1925. And Einstein? He found himself grappling with an uncomfortable consequence of his own discovery. The indistinguishability of particles clashed with his unwavering belief in an underlying classical theory. Once again, Einstein was staring his own spooky quantum physics straight in the face . . .

To conclude this section on bosons, let's revisit the Herculean cooling process required to make a Bose–Einstein condensate. If the soup's too hot, you simply set the bowl on the table and wait for it to cool down. Blowing is allowed (provided Mum isn't watching). During this cooling process, the hottest molecules evaporate until the soup reaches room temperature (a process aptly called 'evaporative cooling'). But cooling a system down to the extremes required for a Bose–Einstein condensate takes more than a patient wait and a puff of air. For that, we need lasers.

THE LASER

The laser, short for 'light amplification by stimulated emission of radiation', is a quintessential offspring of quantum, making its debut in 1960. Since then, lasers have propagated across the globe, becoming indispensable to modern life. Imagine a world without them, and you'd see society and information technology crumble. Lasers transmit data via glass fibres and the internet, extract music from CDs, cut metal sheets with sub-millimetre precision, scan texts and drawings, and erase regrettable tattoos of ex-lovers' names. In the medical world, they remove malignant tumours, work miracles in ophthalmology and even resolve flaws in our blood vessels.

The notion that something like a laser could exist was a hunch of – who else? – Albert Einstein. While some people unwind with puzzles or pull weeds from around the tomatoes, Einstein, as a side gig to revolutionizing physics with relativity, dabbled in pondering other deeply tricky ideas. Inspired by Planck's radiation law, he predicted as early as 1917 that a process of 'stimulated emission' could cause light to be amplified in a coherent way, producing an extraordinarily precise and powerful beam that neither converges nor diverges.

A quick rundown of how a laser works. Under normal circumstances, an electron jumps to a higher energy level when it absorbs a photon with precisely the right energy. When it drops back to the ground state, it emits a photon to compensate for the loss of energy. Lasers take this further: they coax as many electrons as possible into a higher energy orbital, so they can all drop down collectively. Ordinarily, an electron might linger in its excited state – perhaps enjoying a leisurely tea break on the observation deck. But when a photon with precisely the right energy comes along, it hastens the electron's return trip (we'll happily spare you the rather complicated logic behind this). Then there's not even time to order a cup of tea. Now, there are two photons. These identical photons, in turn, encourage other electrons to drop, triggering a chain reaction. Because all the photons are perfectly in sync (sharing the same phase, frequency, polarization and direction) their effect is exponentially amplified. This is where the term 'stimulated emission' comes from.

In addition, an ingenious system of mirrors traps the light within the laser, bouncing it back and forth to create an even greater cascade of photons, each reinforcing the others. In some way, lasers resemble one big Bose–Einstein condensate of light particles: the light particles are all identical. The difference is that in a laser, these particles are in an excited state and not in the ground state, so there's no need to chill the system to near-absolute zero.

This is Einstein at his finest. And what makes it even more fascinating is that with the laser, Einstein invented the instrument that would make it possible, albeit decades later, to put his own Bose–Einstein condensate into practice.

IN A NUTSHELL

> Rutherford cracks the atomic nucleus and breathes life into nuclear physics.

> High time to gauge your theories: may the forces of the standard model be with you.

> A little ray of sunshine: I exist, therefore I am a star.

> *Dramatis personae*: Marie Curie, Ernest Rutherford, Enrico Fermi, Lise Meitner, Robert Oppenheimer, Maria Goeppert Mayer, Hideki Yukawa, Richard Feynman, Shin'ichirō Tomonaga, Julian Schwinger, Chen Ning Yang, Robert L. Mills, Philip W. Anderson, Peter Higgs, Robert Brout, François Englert, Gerardus 't Hooft, Murray Gell-Mann, Steven Weinberg, George Gamow, Fred Hoyle.

SEVEN

PUDDING AND QUARK

7.1 Subatomic physics: the experiments

It is harder to crack prejudice than an atom.
— ALBERT EINSTEIN

With nuclear physics, we're quite literally coming to the core of the quantum story. As seen in previous chapters, physics' greatest discoveries emerged from a dynamic interplay between experiment and theory. Nuclear physics followed much the same trajectory, forged through relentless searching, finding and not finding, and starting over. Let's begin with the experiments.

In 1896, Wilhelm Röntgen discovered a form of radiation so powerful that it allowed him to (literally) see through his wife's hand. Using these mysterious rays, he captured an extraordinary image of her bone structure in unprecedented detail. As the rays were entirely unknown to science, he dubbed them X-rays.

Henri Becquerel (1852–1908) was deeply impressed by Röntgen's work. Just a few months later, he stumbled upon a discovery of his own – *purement par hasard*. Becquerel was convinced he could produce X-rays by exposing phosphorescent uranium salt crystals to sunlight. He believed that the uranium salt would slowly release its stored solar energy in the form of X-rays. However, as often happens, life got in the way, and he tucked the uranium salts into a desk drawer, forgetting all about them. Until one run-of-the-mill day a few weeks later, when Becquerel opened the drawer and was astonished to find that the uranium was still emitting radiation. His hypothesis was completely wrong! The radiation hadn't diminished one bit. Heating, cooling, smashing the crystals into fragments, or even hurling them at the ceiling – all these antics failed to alter the radiation's intensity. It became clear that this mysterious radiation wasn't tied to sunlight or chemical bonding, but originated deep within the uranium itself. Becquerel knew he was onto something and decided to hand over the enigma to others for further investigation. And so, it ended up in the hands of Marie and Pierre Curie.

MARIE CURIE

Madame Curie (1867–1934) was a woman of many firsts: the first in France to earn a doctorate, the first female professor and the first woman to be laid to rest in the Paris Panthéon. She attended the first Solvay Conference and was the only person to participate in seven consecutive editions. Most remarkably, she became the first scientist to win Nobel Prizes in two distinct fields (physics and chemistry). By standing on the shoulders of her predecessors and relentlessly pursuing her belief that there is more to be discovered, she left an indelible legacy in the sciences. Thanks to Marie Curie, the twentieth century marched confidently toward the quantum revolution.

Marie Curie

In 1894, Marie met Pierre Curie, the man whose surname she would adopt instead of her Polish Skłodowska. Initially, she resisted marriage, fearing it would derail her plan to return to Poland after completing her studies in Paris, where she had ventured with singular determination to study physics. But Pierre, unluckily for her plans but lucky for her heart, was not only a physicist but also a passionate poet. The eloquence of his love letters won her over, leading to a shared life that, when they weren't immersed in work, was joyfully spent cycling through the countryside.

Pierre had precision instruments, Marie an unrelenting curiosity. Together, in their makeshift little laboratory, this marriage of minds confirmed Becquerel's hypothesis: uranium emitted its mysterious radiation all by itself. More broadly, they demonstrated that certain materials could release extraordinary amounts of energy with remarkable efficiency, all without combusting. The Curies named the phenomenon radioactivity and embarked on an assiduous quest for other minerals with similar, or perhaps even greater, luminous allure. Their efforts paid off spectacularly: the Curies discovered polonium (named after Marie's homeland) and radium (described by the press, with quintessential French flair, as '*un métal conjugal*'), which emitted a million times more radiation than uranium.

The romantic duo couldn't get enough of that mysterious, enchanting glow. Marie even kept a small vial of radium on her nightstand. Tragically, this fascination had dire consequences. She died of leukaemia, a result of prolonged exposure to uranium and radium. Monsieur Curie, meanwhile, met his end not through radiation but in a tragic accident involving a horse and cart. Fermi and von Neumann, both alarmingly lax in their handling of radioactive materials, later also succumbed to cancer, as did countless young women, the so-called Radium Girls, who painted watch dials with radium-laced paint in poorly ventilated factories across the US. Ironically, Curie's work revealed that radium could be used in cancer treatment. To this day, Marie's cookbooks, along with her research notebooks, remain tainted with traces of highly radioactive – and decidedly unpalatable – radium.

The Curies were acutely aware of the precarious implications of their discoveries. Concluding his Nobel Prize acceptance speech in 1903, Pierre Curie delivered a stark warning that was echoed decades later in Einstein's 1939 letter to President Roosevelt; a caution about the dangers of uranium, not only for public health, but also for national security. What if this potent substance fell into the wrong hands? Marie Curie herself wondered whether humanity was truly better off unlocking all of nature's secrets, given the reckless acts of vengeance and greed that we are often given to indulge in.

Why does Marie Curie deserve a place in the pantheon of quantum scientists? Because her experiments led directly to the key question: what, exactly, are atoms made of? What is their structure? And how can radioactivity be explained as a process that creates entirely new particles? Moreover, she became the chief supplier of radioactive materials to Ernest Rutherford, the outspoken New Zealand physicist who famously turned the field upside down. Through her pioneering work, she laid the groundwork for the most important discoveries in subatomic physics.

ERNEST RUTHERFORD

Ernest Rutherford

Ernest Rutherford (1871–1937), whom we encountered in Chapter 3, was probably the world's greatest experimentalist ever. The fact that he was laid to rest beside Newton in Westminster Abbey speaks volumes. Rutherford had an uncanny ability to devise exactly the right experiment to tackle pressing questions – or even to raise questions no one had yet thought to ask. And he didn't need extravagant equipment in order to do that. Hand him an ordinary roll of aluminium foil, and he would arrive at insights became ever more significant as the years rolled by.

For one of his earliest experiments, conducted at the turn of the twentieth century, he sought to investigate the enigmatic glow emitted by uranium. Rutherford wrapped some uranium in foil and discovered that a few layers were sufficient to stop one particular type of radioactive ray. Adding additional layers stopped another type of ray, leaving only a third kind of radiation that could still pass through. It would take two decades for this discovery to be properly interpreted, but Rutherford was already confident enough to declare a foundational truth. There are three types of radiation: alpha rays (α), beta rays (β) and gamma rays (γ). A subsequent set of ingenious experiments unmasked their identities. Alpha rays are helium atoms deprived of their electrons, the beta rays turned out to be electrons, while the gamma rays are electromagnetic waves. The truth, as Rutherford discovered, was a simple, unembellished pudding – no currants, just the bare essentials of atomic insight.

The Discovery of the Atomic Nucleus

When Rutherford and his students, Hans Geiger and Ernest Marsden, took to experimenting in 1909, the stage was set to dismantle Thomson's plum pudding model. The trio bombarded an ultra-thin sheet of gold foil with alpha particles, expecting them to pass through unimpeded, or at most be slightly deflected ('scattered') by the currant-like negative electrons embedded in the positive 'pudding'. But to everyone's amazement, some alpha particles bounced straight back, as though they had struck something solid. Rutherford's conclusion came to him out of thin air, and was as startling as it was revolutionary: the nucleus of the atom is extraordinarily small compared to the atom's overall size, yet it contains nearly all of the mass. An atom, he realized, consists of a very small nucleus, while the rest is empty space, except for the electrons. Electrons, he imagined, are like flies in a cathedral, buzzing far away from the nucleus, at a distance corresponding to the size of the atom. If an alpha particle happens to hit the nucleus (a rare event) it bounces straight back.

As we explored in Chapter 3, this atomic model was a seismic revelation, shaking the foundations of science. Few discoveries could have been more disruptive at the time. Rutherford's discovery not only inspired Bohr to create his own atomic model with quantized orbitals, but also paved the way for our understanding of the nucleus, whether concerning its composition in terms of protons and neutrons, nuclear fission and the strong and weak nuclear forces.

The First Artificial Nuclear Reaction and the Discovery of the Proton

As the atomic Matryoshka doll continued to be taken apart, it yielded further surprises. In 1919, Rutherford carried out the first-ever artificial nuclear reaction. To do this, he had bombarded nitrogen atoms with his 'trusted right hand': alpha particles. The result: the alpha particles were absorbed into the nucleus, causing it to eject a positively charged subatomic particle. What remained was an oxygen atom. For the sake of historical accuracy, it would take another six years before Patrick Blackett, one of Rutherford's students, correctly interpreted this experiment.

For Rutherford, identifying the mysterious subatomic particle was child's play. It was a hydrogen atom stripped of its electron, later renamed the proton. This proton was emitted as a result of a nuclear reaction, a mutation in the atomic nucleus that led to the birth of a different atom. Rutherford hypothesized that this new atom could, in turn, break apart into a lighter atom and another proton, repeating the process until only protons remained. His conclusion? All atoms in the universe are made from the same fundamental building blocks: protons and electrons!

With this experiment, the alchemists and *diggi-daggi-schurry-*

murry-magicians were proved right after all.[1] Ever since the Middle Ages, alchemists had searched in vain for the Philosopher's Stone that could transform ordinary metals into precious metals – ideally gold. Even Newton, the last of the magicians, had devoted a significant part of his life to this pursuit. Rutherford was the first to succeed.

Mendeleev's conjecture – that all matter on Earth is composed of a few dozen distinct atoms – was already an extraordinary insight. Rutherford's claim – that all atoms are made from the same tiny fundamental components – might *seem* only a small step forward, but in reality, it was a giant leap towards the reductionist view of nature. All atoms are made up of the same tiny Lego bricks. And if we could understand how to reconstruct these bricks, we might unravel even more of nature's deepest mysteries.

Rutherford, ever the determined investigator, continued his quest to unlock the secrets of the nucleus. It couldn't just contain protons – something else had to be rattling around in there. The nucleus was simply too heavy. By 1920, Rutherford was convinced that another kind of particle, essential for holding the whole system in equilibrium, had to exist.

The Discovery of the Neutron

It wasn't until 1932 that yet another of Rutherford's students at Cambridge, James Chadwick, finally provided experimental proof that Rutherford was right. Chadwick had found a particle with no charge – neither positive nor negative. Neither? Does it have chemical properties, then? No . . . How so? And does it have a mass? Yes! It has a mass! But what is it, exactly? Hmm, hard to say . . .

---

1 A reference to Mozart's *Bastien und Bastienne*.

Hold on, let's fire some alpha particles at the nucleus! And so, with unwavering determination, alpha particles – along with beryllium atoms – were fired at the nucleus. And? Incredible but true: a neutral mystery particle emerged, with a mass just a smidge more than that of the proton![1] So . . . they're essentially neutral protons? Great, let's call them neutrons! Hence Rutherford and Chadwick not only identified the neutron, but also gave it a name. As 1932 drew to a close, the last fundamental building block of every atom had finally been discovered.

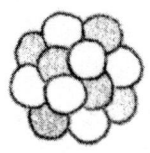

The atomic nucleus is a dense cluster of protons and neutrons.

With the discovery of this new particle, the neutron, it became clear why Mendeleev's table needed an extra dimension. Some atoms had the same number of protons/electrons, but different weights. These are isotopes, a word derived from the Greek *isos* (equal) and *topos* (place). The difference lies in the number of neutrons. Take carbon, for example: in nature, we find two isotopes of carbon, each with six protons. The first isotope (C_{12}, or carbon-12) contains six neutrons, while the other isotope (C_{14}) contains eight neutrons (the numbers 12 and 14 refer to the atom's mass, or rather: the sum of its protons and neutrons). The missing link in unravelling the nucleus was precisely the realization that different isotopes exist, and that

---

[1] We now know that neutrons in themselves are not stable. After fifteen minutes, they decay into a proton, an electron and a neutrino. This is why they weigh slightly more than protons. But they *are* stable in an atomic nucleus, due to interactions with the other nucleons.

their defining feature is the number of neutrons in the nucleus. This revelation opened up an entirely new world of possibilities and forced Rutherford to think deeply . . .

The Age of the Earth

. . . about the Earth, for example. How old is the Earth, really? With the discovery of isotopes, it finally became possible to work this out. Lord Kelvin had previously tried to estimate the Earth's age, based on the temperature difference between its core and its surface. His figures ranged between 20 million and 400 million years. Rutherford, however, saw room for far greater precision.

He suspected that the Earth had accumulated many more grey hairs within the folds of its crust and devised a far more accurate method of calculation. First of all: radioactive substances take time to transmute into other substances. But not all of them do so at the same rate. Why not? Because this process is governed by quantum mechanical tunnelling, which is completely random. Every radioactive substance decays at its own pace. Since this process can take a very different amount of time, Rutherford introduced the term 'half-life': the average time it takes for half the atoms in a radioactive material to be 'reincarnated' in another element. In the next equivalent time span, the remaining amount will halve again. Take uranium-238: it's nothing if not stubborn. It takes 4.5 billion years for half of its atoms to decay and turn into radium-226. Radium 226, by contrast, completes its transformation to the noble gas radon-222 in a mere 1,600 years. This explains Marie Curie's discovery that radium-226 produces a million times more radiation than uranium-238.

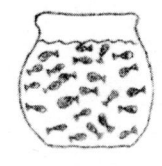

Half-life explained: imagine you have a bowl with twenty-four fish, but each night half of them are mysteriously eaten. On day two, only twelve fish remain. On day three, just six are left. On day four, that number drops to three.

When the Earth was still in its infancy, its molten surface began to cool, forming solid rocks. Deep within these rocks, uranium was locked away as the searing, flowing slurry solidified. Over time, uranium decays, ejecting other atoms that also become trapped within the rock. After a few fleeting millions of years or so, some rocks manage to break free and are brought to the surface by a volcanic eruption, an earthquake, tectonic plate shifts, a human digging a well, mining, or other such down-to-earth activities. And so, ancient rocks found their way into the hands of young brainiacs like Rutherford, determined to find out how long this spinning sphere of ours has been around. A bit of scientific poking and prodding revealed that uranium-rich rocks also contained traces of helium and lead. The presence of lead in a rock was a sure sign that it had been aging for quite some time.

Long story short: by measuring the concentrations of specific elements, factoring in their half-life, scrutinizing the isotopes and applying a few other complex yet logical formulas, scientists could estimate how much material a rock originally contained at its 'birth'. By comparing these results to what remains today – and in what form – it turned out that the Earth had to be around 4.5 billion years old (the Big Bang, for context, dates back to 13.8 billion years ago). There is no longer any real doubt about this. Unless you go to school in Wisconsin.

Can the same technique be used to determine the age of the Turin Shroud? Absolutely – provided we look at isotopes that decay a lot faster. For billions of years, all life on Earth has been sustained by the heat (electromagnetic radiation) generated by nuclear reactions in the sun. But the sun also churns out cosmic rays, which have crazy amounts of energy and consist mainly of atoms stripped of their electrons. Our atmosphere is constantly bombarded by cosmic rays. These particles head towards Earth at nearly the speed of light. Fortunately, the radiation is absorbed by our atmosphere, triggering a fascinating nuclear reaction: the cosmic rays convert nitrogen and neutrons into carbon and protons (n + N_{14} -> C_{14} + H). Like all living organisms, our bodies are packed with carbon absorbed from the atmosphere – mainly C_{12}, with a tiny fraction (one trillionth) of C_{14}. Considering that carbon accounts for around 23 per cent of our bodyweight, that tiny fraction of C_{14} is still surprisingly significant. Carbon is the elementary building block of life – emphasis on 'life'. Because when living organisms die, they stop absorbing fresh C_{14} from the atmosphere. Interestingly, C_{14} is not stable. Over time it decays back into N_{14} (with a half-life of 5,730 years). So in the absence of respiration, the C_{14} stored in humans, trees or plants, begins to decay, while the stable C_{12} remains unaffected. By measuring the ratio of C_{14} relative to C_{12} in a lifeless body or a felled tree, we can estimate how old something was when it died (or: how long it has been since it stopped absorbing carbon). The Turin Shroud, made of plant-derived fibres, follows the same rules. Once the plants were harvested, their C_{14} began to decay in the same way. By applying this technique, three independent research centres dated the Shroud to between the mid-thirteenth and late fourteenth century. Doubts and scepticism aside, carbon dating remains one of the most precise and endlessly captivating tools for exploration.

The First Controlled Nuclear Reaction

Brilliant. But Rutherford wasn't done sleuthing yet. His gut feeling told him there was still more to uncover in the nucleus. In 1932, working with two fresh-faced students, Ernest Walton and John Cockcroft, he achieved the first controlled nuclear reaction. Using a particle accelerator, the team bombarded lithium with protons. The lithium was converted into alpha particles and (ta-da!) a tremendous amount of energy was released. The latter effect was unprecedented. Except: keeping the particle accelerator running required far more energy that it produced – not exactly efficient. But not a waste of effort either, because the experiment had revealed all sorts of things about the atom. No one at the time could have imagined that, a decade or so later, this knowledge would lead to a technology that produces colossal amounts of energy . . .

To be clear, the quest to understand nuclear reactions was never about finding powerful new sources of energy. Scientists like Rutherford were, above all, driven by an insatiable curiosity, daring to explore the hidden depths of the tiniest particles. They sensed that monumental breakthroughs still lay waiting within the atomic nucleus – that connecting one puzzling phenomenon to a bold new interpretation could lead to groundbreaking discoveries. These researchers sought to dismantle the intricate Matryoshka doll of nature, unravelling the atom and its ever-smaller elementary particles.

Before we bid Rutherford farewell, one final thought. He provided yet another clear demonstration that intuition is perhaps the single most important instrument at a scientist's disposal. Rutherford moved mountains of work, had the energy of an alpha particle and was unrivalled in arriving at one great insight after another. The older he got, however, physics itself seemed to grow younger. Despite his characteristic diligence, he watched enviously as experiments grew ever more complex, equipment ever more sophisticated and theory ever more abstract – while he himself became increasingly isolated in his laboratory. With the unstoppable rise of mavericks like Einstein, Heisenberg and Schrödinger, physics was undergoing an inevitable phase transition. But before we move onto the theory of nuclear physics, let's linger a bit longer in the experimental lab.

ENRICO FERMI

> *Mysterious, even in broad daylight,*
> *Nature won't let her veil be raised:*
> *What your spirit can't bring to sight,*
> *Won't by screws and levers be displayed.*
>
> — WOLFGANG VON GOETHE, *FAUST*

In the wonderland of the tiniest things, a new toy had captured everyone's attention: the neutron. It was like a pinball ricocheting through the machine of nuclear physics. Neutrons, having no charge, are not diverted by electromagnetic forces, they barely interact with electrons, and remain unfazed by protons. Unperturbed, they burrow blithely through matter. At the same time, these properties make them ideal for the further study of the inner structure of atoms and materials. That hadn't escaped Fermi's notice. Except it's one thing to establish that they have such penetrating power; knowing *why* neutrons can penetrate into matter so much deeper than, say, alpha or beta particles, that's another thing altogether.

Meanwhile, in the Paris of 1934, Irène Joliot-Curie (yes, the daughter) and her husband Frédéric Joliot had discovered that if you bombard boron, phosphorus, aluminium or magnesium with alpha particles, you can artificially create radioactive elements. Not bad, thought Fermi, but surely it could be improved. *Andrà meglio con dei neutroni!*[1] And so, he redid the experiment with neutrons instead of alpha particles.

Enrico Fermi

Through much trial and error, and guided by his razor-sharp intuition, Fermi discovered that slower neutrons interact much more strongly with matter than fast neutrons. But first, he needed to find a way to slow them down. How do you decelerate something as elusive and intangible as a neutron, without bringing it to a complete halt? After all, it wasn't as if he could just pluck an instruction manual from his bookcase to find out what he was supposed to do. He had once casually suggested that paraffin might do the trick – and, remarkably, it did. This meant he could now use neutrons to bombard uranium much more efficiently. The results were, to put it mildly, profoundly intriguing. Fermi was awarded the Nobel Prize for his work on slow neutrons and the discovery of new nuclear

---

1 'It will go better with neutrons!'

reactions – even though, as it turned out later, his experiment had been fundamentally misunderstood, not only by the Nobel Committee but also by Fermi himself. No one realized that he had inadvertently achieved artificial nuclear fission for the first time. For clarity: nuclear fission, unlike nuclear fusion (where lighter elements fuse to form a heavier one), involves splitting a heavy element into lighter ones, a process that also unleashes an extraordinary amount of energy.

LISE MEITNER AND OTTO HAHN

Lise Meitner

Lise Meitner (1878–1968) and Otto Hahn (1879–1968) were a scientific dream team: she the sharp-minded theoretician, he the meticulous experimentalist. With their signature German precision, they had a few reservations about Fermi's spirited yet somewhat slapdash experimental methods. Meitner and Hahn insisted on analysing everything down to the smallest detail, measuring and understanding every step. Fermi may have been a formidable theorist, but on an experimental level he sometimes leaned towards a quick-and-dirty approach. So, where were the cracks in his work?

In 1939, after years of painstaking research, and aided by the chemist Fritz Strassmann, the pair reached an astonishing conclusion: Fermi's experiment had, in fact, triggered nuclear fission. They observed, with unambiguous clarity, that bombarding uranium with slow neutrons, causes the atom to 'split' into lighter nuclei (such as barium), along with a cascade of alpha particles, beta particles and neutrons. And that 'lost' mass? Not lost at all! It was converted into energy. Importantly, during nuclear fission more neutrons are released than are fired in. These surplus neutrons can then split other uranium atoms, triggering more neutrons in a self-perpetuating chain reaction. The number of neutrons therefore increases exponentially, culminating in a staggering release of energy. Fermi hadn't looked at it that way – what you don't expect to see, you often overlook, even if you're as brilliant as he was. But the result was unmistakable. The implications hit like a bombshell, quite literally. Meitner and Hahn's insights into nuclear fission opened the door to the possibility of a highly explosive powerful weapon. Oh dear.

At first glance, the partnership between Meitner and Hahn seemed impeccable; he searched, she deciphered. He experimented, she interpreted. Meitner brought the sharp analytical mind that Hahn lacked, while Hahn's deep knowledge of chemistry provided the indispensable backbone that ensured their experiments succeeded and tallied with the theory. Their partnership was not so much amorous as rigorous. They took great care to keep their working relationship free from the toxic gossip of academic corridors. When Austria was annexed by Germany, Meitner increasingly felt the weight of her Jewish heritage. The Berlin institute where she was working suddenly deemed her dangerous. Her expertise and analytical prowess were no match for the prevailing ideology. She was shown the door. Hahn, who had promised to help her, eventually left her to fend for herself. Was it cowardice? Pragmatism? The truth likely lies somewhere in between. Meitner ended up in Sweden, a safe but isolating haven, where she lacked access to a proper laboratory. Yet her desire to work never wavered. Correspondence with Hahn resumed, albeit reluctantly. He conducted the experiments, sending her the results, and she replied with her ever-detailed interpretations. Through this chain of letters, the foundational discoveries were made in the field of nuclear chain reactions.

When Hahn, along with Fritz Strassmann, received the Nobel Prize for the latter discovery, he didn't mention Meitner once. Not a word about their collaboration. Not a word about their letters. Not a word about her critical contributions. Yet without her, he would never have reached such heights. Everyone agreed that she deserved to share that Nobel Prize. But Meitner, far from vindictive, let her disappointment subside as it became ever clearer how the splitting of uranium could – and would – be misused. The less her name was tied to the inevitable bomb, the better. This 'German Marie Curie', as Einstein affectionately called her, was never to set foot again in Germany. No planet was named after her. Her legacy was immortalized in another way: element 109 in the periodic table, meitnerium, a highly radioactive element that rapidly decays into bohrium, named after her fellow luminary, Niels Bohr.

Einstein famously doubted that atomic energy would ever find practical applications. Little did he know that the coming decades would be dominated by nuclear reactors, nuclear power and atomic bombs. Because by now the 1940s were approaching. At a geopolitical level, the world was in crisis and ominous reports were pouring in from Germany.

ROBERT OPPENHEIMER

There was once an American physicist with two left hands. Lab work? Not for him, he was advised. Experimental physics was best left to those with steadier grips and surer instincts. So, he set his sights on theoretical physics instead. His admission to Christ's College in Cambridge carried the promise of a foolproof future and a welcome distraction from his gloomy mood (his romance had just fallen apart and his relationship with the other significant woman in his life, his mother, was strained, to say, the least). At Cambridge, the young student found himself under the supervision of a professor who excelled in the laboratory and insisted that his protégé would also get his hands dirty with experiments. This demand awoke the wicked witch within the student, who hatched a rather dramatic plan, or so the myth proclaims. He left a poisoned apple on the desk of his despised Snow White, Patrick Blackett (Rutherford's aforementioned student). Having enacted his prank, he fled for a holiday with friends to let off steam. Big-mouths being what they are, he couldn't resist bragging about his bravado, and his companions wasted no time in reporting the incident to the university authorities. The administration leapt into action, suspending the student with the polite suggestion to kindly report to the nearest mental institution as soon as possible.

Patrick Blackett lived happily ever after, going on to discover cosmic rays. Robert Oppenheimer (1904–1967) – yes, the very same student – also went on to have a long and illustrious career, though not at Cambridge. Max Born had brought him to Göttingen. There, 'Oppie' (as his friends called him) cemented his reputation as one of the world's most notorious and controversial physicists of his time. It was there, with Born, that he became the first to understand (and explain) the structure of molecules in terms of quantum physics.

NIM NIM

Emmy Noether had her Noether boys, and years later in Berkeley, Oppenheimer had his 'nim nim' boys. His students were utterly captivated by their charismatic mentor, to the point of imitating his every quirk. They chain-smoked the same Chesterfields, walked like him, talked like him, shared his taste for sharp suits, echoed his distinct accent and mimicked the peculiar 'nim nim' noises he made when lost in thought.

Oppenheimer grasped the quantum behaviour of molecules and quantum fields like no other. But he also liked to look up. Oppie laid the foundations for all future research into black holes and cosmic rays. Across the United States, there wasn't a quantum physicist of his calibre. Yet with such brilliance and potential, it was inevitable that he would catch the eye of the US military. Under their auspices, he rose to become the linchpin of one of the United States' most ambitious and secretive endeavours, the Manhattan Project. This project had one singular goal: to develop the atom bomb. The project's success owed much to its unprecedented assembly of the world's brightest minds.

After the pioneering work of Meitner and Hahn, the Americans soon realized that Germany would waste no time in building its own atom bomb. Ironically, it was a 1933 Nazi law banning Jews from public office that sent waves of Germany's Jewish *éminences grises*, including professors of physics, across the Atlantic. Many of these expatriates contributed to the Manhattan Project. Germany had shot itself pretty accurately in the foot with this law, because in the end it was not Nazi Germany but the Allies that produced the first atom bomb. It is nothing short of miraculous that, just two years (and $2 billion) later, such a device was developed, given that, from an engineering perspective, it remains one of the most complex undertakings ever achieved in history.

Today, Oppenheimer is largely remembered as the father of the atom bomb. Some take a slightly more constructive view, highlighting his role in hastening the end of the Second World War. Yet Oppenheimer found the use of science as a weapon deeply unsettling. In later years, he advocated for atomic power to be used for peaceful purposes. His story took a tragic turn when he became a high-profile target of Senator McCarthy's anti-communist crusade. Although the atom bomb undeniably confirmed America's position as the world's strongest power, that same nation viciously and mercilessly destroyed the man to whom it owed that status. Oppie fell into a profound depression, even darker than the one he had previously wrestled with. He was brilliant, yes, but not as brilliant as Pauli. He was fairly ingenious, certainly, but not quite on the level of Fermi and Heisenberg. The reason is that Oppenheimer – as Freeman Dyson so expressively put it – had no *Sitzfleisch*, the ability to sit still and do long calculations. That was his glorious flaw. Extended periods of focused work were just not in his nature. Yet, solving the truly tricky problems, of course, is not something you can do on the fly. Unless, of course, your name is Einstein.

CHICAGO PILE-1 AND THE MANHATTAN PROJECT

It all began with a team of scientists centred around Enrico Fermi, who by now had become a professor in Chicago. The group had built a lab out of 'a crude pile of black bricks and wooden timbers' and transported tonnes of graphite and uranium[1] into the disused space under the stands at the university's football field. Boundless confidence was shown in Fermi and his ability to complete this ultra-secret mission – apparently not even his wife knew about it. Fermi and his team did something no one had ever done before. In 1942, they initiated a controlled nuclear chain reaction. In other words, Chicago was the site of the first-ever nuclear reactor. This landmark achievement proved that nuclear power was not just theoretical – Einstein's scepticism notwithstanding – but could also lead to the construction of a bomb. And that's when things really began to heat up.

Among the Manhattan Project's contributors were some familiar names: Eugene Wigner, Enrico Fermi, Niels Bohr, Richard Feynman, Hans Bethe. At the helm was Robert Oppenheimer. Albert Einstein was not directly involved. He was not a nuclear physicist and, more importantly, he had little taste for political machinations. In truth, Einstein was deliberately excluded from the project due to his alleged communist sympathies. Indirectly, though, he was a tiny bit involved. It all began with a letter he penned to President Roosevelt, under persistent prodding by (the less influential) Eugene Wigner and Leo Szilard. In this letter, Einstein explicitly referred to Fermi's work, and chiefly warned 'that the element uranium may be turned into a new and important source of energy in

---

1 Over 1,000 tonnes of uranium ore – virtually worthless at the time – had already been shipped from the Belgian Congo to Staten Island. Belgian mining engineer Edgar Sengier, of the Union Minière du Haut Katanga, had a remarkable intuition that the ore could play a significant role in the war effort against Germany. His foresight earned him the distinction of becoming the first non-American civilian to receive the Medal for Merit from the United States government.

the immediate future [. . .] and [. . .] that extremely powerful bombs of a new type may thus be constructed'. Einstein, ever a staunch advocate of liberty and free speech, even reconciled with his nemesis Bohr on this issue, but otherwise preferred to stay on the sidelines. 'Scientists who know how to get a hearing with political leaders,' he argued, 'should bring pressure on the political leaders in their countries in order to bring about an internationalization of military power.' His warning had an unintended consequence: the United States promptly decided to develop its own atomic bomb.

7.2 Subatomic physics: the theory

In the late 1920s, quantum theory was still in its infancy. While many a mind was mulling over Niels Bohr's theories and searching for ways to reconcile atoms and molecules with wave mechanics, George Gamow (1904–1968) chose to look the other way, where it was quieter. What if the atomic nucleus were examined from a different angle? The release of an alpha particle, he posited, might be analogous to the emission of a photon when an electron drops from a higher to a lower energy orbital. Could something similar be happening in the nucleus? Could a particle in the nucleus jump between energy levels in the same way? Gamow's conclusion (or rather, speculation) in 1928 was that radioactivity had to be a purely quantum mechanical phenomenon; within the nucleus, a quantum mechanical process occurs. But instead of electrons, it's the protons and neutrons that are doing the jumping. As they do this, they emit alpha, beta and gamma particles, causing the nucleus to reorganize itself and transform into a new atomic nucleus.

The atomic nucleus had already been chopped up and scrutinized *ad nauseam*, yet despite all the experiments, there was still no clear theory that explained how to define the structure and workings of this tiny treasure. Nuclear orbits, however, do not contain only one type of particle (like the electron for an atomic orbital) but two (neutrons and protons). This makes it far more complicated to describe mathematically than the electron structure in Mendeleev's periodic table. It would take a few more decades, until 1950, for that puzzle to be solved. In that year, Maria Goeppert Mayer (1906–1972) created a periodic table of all the possible proton and neutron orbitals within the nucleus: the nuclear shell model.

Maria Goeppert Mayer was only the second ever woman to receive the Nobel Prize for Physics, in 1963. Like many other scientific giants, she attended the University of Göttingen. And, like many other female intellects of the time, she systematically missed out on academic appointments, finding herself relegated to unpaid or poorly paid jobs. But that didn't harm her talent. She had arrived at her nuclear shell model by building on the work of Eugene Wigner, who had rooted his own research in symmetry principles and in generalizing the electronic S, P, D and F orbitals. Unsurprisingly, her theory showed many similarities with Bohr's atomic model. Mayer proposed that the nucleus, much like electron orbitals, had to consist of different 'shells' on which protons and neutrons organize themselves to achieve stability. Her diagrams, layered like an onion, earned her the less-than-flattering nickname from the ever-maudlin Pauli: 'Onion Madonna'. Enough to make you cry your eyes out.

Mayer concluded that protons and neutrons are circling and twirling in the nucleus, like a ballroom full of waltzers, chaotically dancing yet flawlessly in sync, their steps dictated by the energy level they occupy. A nucleus containing 2, 8, 20, 28, 50, 82 or 126 protons or neutrons exhibits greater stability, akin to how noble gases achieve stability by completely filling their electron shells. Eugene Wigner fittingly called these the 'magic numbers'.

By the way, Maria Goeppert Mayer shares a celestial honour with Emilie du Châtelet: a crater on Venus. Sigh . . . Let's keep our feet on the ground and dig a bit deeper into the core.

THE WEAK AND THE STRONG NUCLEAR FORCES

But how to explain all of this? There are still two outstanding mysteries. First of all, what force holds the protons together in the nucleus? Second, what force drives the radioactive processes that occur there? The answer: the strong nuclear forces, for the first. The weak nuclear forces, for the second.

It was this second question that kept Fermi up at night, and eventually led to his discovery of the neutrino. Here's how. An atomic nucleus doesn't necessarily have to be bombarded with particles for all sorts of things to happen. A neutron can decay on its own. The weak nuclear force enables this, converting a neutron into a proton. But because the proton is positively charged, the neutron must emit an additional particle to balance things out. Something negative. An electron! However, another problem arose: when scientists measured the energy, the sum of the energies of the proton and electron was less than the energy of the original neutron. The logical conclusion was that another particle must be involved to account for the missing energy. Could it be an extra neutron? No. Out of the question. Too big and too heavy; there's not enough energy for that. Another proton? Impossible – it's positively charged. Perhaps something smaller? Yes, a tiny mini particle! A neutrino, as it is so poetically named in Italian (meaning 'little neutral one'). Neutrinos have hardly any mass, and zero charge, but they do have energy and momentum. Their ghostly nature comes from the fact that they interact with almost nothing, allowing them to pass through everything. They can pass through marrow and bone, through the roof, through underground bunkers and even through lead walls as if they were made of warm wax. We too are constantly being bombarded with neutrinos. Pauli had long suspected the existence of this elusive particle, but it was Fermi who worked out the whole process, called beta decay, and officially introduced this new particle in a mathematical theory.

Where there's smoke, there's fire. If weak forces exist, then surely strong forces must exist as well, mused Hideki Yukawa (1907–1981). On a cosmic scale, gravity holds everything together, but what keeps the atomic nucleus intact? There must be a force that reconciles the apparent chaos, something that counteracts the electromagnetic force pushing protons apart. What is this peacemaker that prevents protons from flying off in all directions? In 1949, Yukawa postulated that all particles in the nucleus (protons and neutrons) attract each other by exchanging new types of particles, which he called 'mesons' – a nod to their mass, falling somewhere 'in the middle' between an electron and a proton. Unlike electrons and protons, however, mesons are not fermions but bosons.

Just as the electromagnetic force (or Coulomb force) between charged particles arises from the exchange of photons (bosons), the protons and neutrons (jointly: 'nucleons') exchange mesons. But there is one fundamental difference: photons do not have a mass, which allows the Coulomb force to stretch over vast distances. Mesons, in contrast, do have mass. As a result, their corresponding forces work only over very short distances: the size of an atomic nucleus. When a proton and a neutron exchange a meson, the process must overcome a significant energy barrier, since the meson has a mass ($E = mc^2$). But nature doesn't like to waste energy. So it 'borrows' the required energy from the vacuum, ensuring it is returned quickly – giving the force a very short range. Particles are eager to pay their energy debts as quickly as possible, meaning the distance between them must remain as small as possible. Conclusion: the more energy needed to create a particle, the faster that energy is repaid, and the shorter the range of the resulting force.

Yukawa's theory explained an awful lot of things about the forces that hold the atomic nucleus together. By predicting the existence of mesons, it ushered in a new era in experimental particle physics and led to the construction of particle accelerators, including those at CERN. Although mesons didn't fully solve the enigma of the atomic nucleus, and Yukawa's theory was still far from complete, it was revolutionary enough to earn him the Nobel Prize for Physics, making him the first Japanese laureate in the field.

SWISS PARTICLES

If you've ever seen photos of CERN, you might wonder whether this gigantic underground doughnut, with a circumference of almost 17 miles (more than 27 kilometres), isn't just a tiny bit disproportionately large for research into particles that are so disproportionately small and light. Not really. CERN is actually the world's biggest microscope. Since the three types of bosons discovered there (more on this shortly) have a mass, and as Einstein taught us that mass is equal to energy (so the heavier a particle, the more energy it takes to create it), results can only be achieved by generating vast amounts of energy. And this simply isn't possible with tiny small-scale machinery.

In a particle accelerator, atoms are whirled around the ring thousands of times per second – a single beam of particles carries as much energy as a roaring freight train hurtling along at hundreds of miles per hour. These atoms are then smashed into each other at near-light speeds, breaking apart into even smaller particles. It's through these violent collisions that scientists can, in peace and quiet, study the tiniest of the tiny particles. And because these splinters appear and disappear in the blink of an eye, you better stay alert and keep your eyes wide open. One particularly remarkable side benefit of the particle accelerator is that it was the birthplace of the World Wide Web. The staggering amount of data produced by these experiments demanded an efficient system for sharing information across multiple computers. Hence, the internet was born.

FIELD THEORIES AND GAUGE THEORIES

Nobody truly understands the conjuring act of infinities, symmetries and gauge bosons that follows. But if you endure their presence for long enough, you somehow learn to live with them. Field theory is an integral part of the quantum canon, so it cannot – under any circumstances! – be omitted here. Still, don't let it tie your brain in knots.

In the post-war years of the twentieth century, yet another revolution took place. With the invention of quantum electrodynamics (QED), the electromagnetic force, quantum mechanics and the special theory of relativity were brought together in a single common, coherent field theory. After all, theories must play nicely together; they're not supposed to contradict each other. The main problem here was the many-particle problem, which could only be tackled 'by approximation', that is to say, using perturbation theory, a method of gradually adding interactions to a Hamiltonian that initially describes non-interacting particles. So why hadn't anyone cracked this nut before 1948, despite the fervent efforts of Heisenberg, Pauli, Dirac, Jordan and Wigner, alias the pioneers of the field? The culprit was perturbation theory itself, which kept churning out infinite values. The solution finally came with the technique of renormalization. This method balances the infinities by allowing the terms in the Hamiltonian to grow infinitely large themselves. The result: a final outcome that is finite, and therefore usable (and measurable). But this mathematical sleight of hand came at a cost: intuition had to step aside, making room for the abstraction of pure mathematics.

This was the moment when the old guard, led by Einstein (Rutherford had already passed away), threw in the towel – maybe just like you're about to. But for the aficionados, we are pressing on for another thirteen pages, because this is without doubt the most complete and accurate framework for tackling the physics of the very small. For those finding this a bit too knotty, we'll gladly see you again in Chapter 7.3.

Here we go. In a field theory, space is represented as a plane filled with points, each carrying its own quantum degrees of freedom – bits of information about the presence or absence of particles, their spin and other properties. This is the only way to construct a theory that harmonizes the principles of the theory of relativity with those of quantum mechanics. Within this space, only local interactions are permitted, meaning particles can only influence their infinitesimally close neighbours. This ensures that information respects the universal speed limit: nothing travels faster than light.

An example of a classical field theory is the physics describing a vibrating string. Here, the degrees of freedom are the string's displacements (how much it moves up or down). The forces at work between the adjacent particles appear as the tension of the string. A quantum field theory is, in essence, a superposition of such classical strings, a vibrating symphony of probabilities.

QUANTUM ELECTRODYNAMICS

In quantum electrodynamics, we focus on how electrically charged particles, such as electrons and positrons (the positive antielectrons, or counterparts of the electron, introduced by Dirac), interact with each other. Since interactions cannot have an instantaneous effect at long distances, there must (again) be another particle mediating the interaction. There must be a particle exchanged between the electrons and/or positrons, a particle that travels no faster than the speed of light and is responsible for the mutual attraction or repulsion. Take a wild guess, what is it? That's right: it's a photon!

The 'field' in field theory is a photon field, where photons are constantly being created and annihilated. Electrons and positrons can also annihilate each other, transforming into a photon with very high energy. In the 'ordinary' quantum physics of Schrödinger and Heisenberg, the number of particles cannot change. Such processes cannot be described. This 'every-particle-for-itself' approach simply doesn't work here. In order to be consistent, the theory *must* take account of a variable number of particles. This is achieved by introducing not only a photon field but also electron and positron fields.

An essential property in this process is that the total charge is always preserved. This constant represents the global symmetry of the system. The photon field ensures that this global symmetry is fully expressed locally as well. In technical jargon, the photon field is a 'gauge field'. As a result, the symmetry group expands, becoming much larger and more interesting. This brings us back to the solace physics offers: symmetry. Gauge symmetries will serve as the guiding principle in the further exploration and understanding of nuclear forces. The deeper we delve into quantum physics, the more we realize just how fundamental Emmy Noether's work truly is.

Back to field theory. The burning question is: what happens in a truly empty space? How do we deal with the vacuum of the system – the 'zero-point or vacuum fluctuations' – since it consists of a big chaotic soup of photons, electrons, and positrons continuously popping in and out of existence? The great challenge in field theory, or rather, in calculating these zero-point fluctuations, is the unavoidable encounter with one of the most slippery concepts in physics: infinity. And let's face it: infinity is always a bit of a touchy subject.

Enter the dream team of Shin'ichirō Tomonaga (1906–1979), Julian Schwinger (1918–1994) and Richard Feynman, joined by a fourth virtuoso Freeman Dyson (1923–2020), who brilliantly demonstrated that the trio's independent findings all boiled down to the same essence. Together, they tackled this conundrum in a way that might make your logical instincts shudder: they assumed that every term in the Hamiltonian was infinite. But here's the kicker: they structured it in such a way that the sum of all these infinite vacuum fluctuations yielded a finite outcome. This mathematical wizardry enabled them to impose a sense of order on the infinite chaos of the vertiginous vacuum.

Their quantum electrodynamics (QED) theory proved itself watertight when the trio made one of the most precise predictions ever seen in the history of physics: the fine-structure constant. The measured energy levels of the hydrogen atom didn't quite align with the quantum physics of Schrödinger and Heisenberg, who hadn't accounted for the zero-point fluctuations caused by interactions with the photon field. QED, however, did factor these in. The theoretical calculations now perfectly matched with the experiments. And this was nothing short of miraculous, all things considered – imagine starting with wild infinities and ending up with a number so precise that it fits neatly on the little screen of your calculator, complete with twelve decimal places, and aligns flawlessly with reality.

In essence, QED demonstrates that, if you tweak the parameters of your theory (such as mass, interaction strengths, etc.), allowing them to become virtually infinite, the final result can still be finite. This process, known as renormalization, essentially involves 'coaxing' infinity into submission – massaging it away, so to speak. It might sound like another mathematical party trick, but it works. And that's why QED is called a renormalizable theory. Feynman, with his 'pathetic little' diagrams, managed to tame infinity like no one else.

YANG AND MILLS

With this success under their belts, theoretical physicists turned their attention to nuclear forces. Could Fermi's weak force, or Yukawa's strong force, also be interpreted as a renormalizable gauge theory? The answer was a resounding no. In both cases, the infinities proved unmanageable, requiring a search for a better, renormalizable theory.

Meanwhile, significant progress was made in unravelling the structure of gauge theories. Pauli had derived his famous spin-statistics theorem, which established a fundamental duality: matter fields (or particles, such as electrons) must be fermions with half-integer spin, while gauge fields (such as photons, which facilitate interactions) must be bosons with integer spin. This is consistent with quantum electrodynamics and, more importantly, it paves the way for formulating gauge theories for the nuclear forces.

Chen Ning Yang (b. 1922) and Robert L. Mills (1927–1999) took this bull by the horns. Their starting point was simple: how can QED be generalized to systems with a much larger, non-commutative gauge symmetry (because in QED the gauge symmetry applied only to charge)? This richer symmetry would then be expected to describe the properties of a whole range of different bosons that organize the forces within the nucleus. On paper, it looked very promising. The problem was that their theory was only consistent if the bosons in question were massless. This was completely at odds with Yukawa's insights, which explained that the strong nuclear force has such a short range precisely because its messenger particles – the mesons – do have mass. Oops. And so, the theory of Yang and Mills was set side.

Surprisingly, inspiration arrived from an entirely different domain: superconductivity (discussed in more detail in Chapter 8). By the mid-twentieth century, a robust theoretical framework for superconductors had emerged, explaining their unique experimental properties. A key figure in this story was Philip W. Anderson, a solid-state physicist who introduced a truly novel idea: in superconductivity, symmetry breaking results in mass acquisition. Specifically, a massless boson (a 'Goldstone boson' resulting from symmetry breaking), when paired with a massless gauge field, can produce a new gauge field *with* mass. Anderson speculated: could the combination of symmetry breaking and a massless gauge field be useful in describing the weak and strong interactions within the nucleus? His insight hinted at a mechanism for giving mass to the bosons responsible for the weak and strong interactions. British physicist Peter Higgs and, independently of Higgs, the American Robert Brout and the Belgian François Englert, picked up Anderson's hint. In 1962, they applied his insights to relativistic gauge field theories. Bingo. Thanks to Anderson's mechanism, the gauge fields of a non-commutative gauge theory could acquire mass. As is often the case in science, the invention was not named after its original architect, but after the one who first applied it successfully. Hence the Higgs mechanism. It is one of the foundations of particle physics.

So, could the Yang and Mills theory still be saved? Absolutely. In 1967, Steven Weinberg and Abdus Salam demonstrated that, with the help of the Higgs mechanism, a Yang-Mills theory could be formulated to unite Fermi's weak nuclear force with Feynman and company's QED. This gauge symmetry was named SU(2)xU(1). Besides the photon, this theory includes three new gauge bosons: the W+, W- and Z bosons. The theory also accounted for a menagerie of hedonic – sorry, *fermionic* matter particles (the leptons, six in total) that had already been discovered experimentally. The most well-known lepton is the electron, which is joined in its category by the less well-known muon, the tau particle, the electron neutrino, the muon neutrino and the tau neutrino (all with corresponding antiparticles). Once again, the guiding hand of symmetry was everywhere to be seen.

The predicted W and Z particles were indeed discovered in 1983 at the CERN particle accelerator. First came the detection of the W bosons, followed a few months later by their neutral counterparts, the Z bosons. Both had masses compatible with Weinberg's theory. By then, theorists had no doubts left about the validity of the weak force theory. Gerard 't Hooft had already demonstrated in 1971, with astounding mathematical virtuosity, that it was renormalizable – just like QED, which was part of the same framework. The remaining question was whether this success could extend to the strong nuclear force. Could the Yang–Mills + Higgs combination provide a definite answer to that as well?

QUANTUM CHROMODYNAMICS

Back to the 1950s. The war was over, society breathed a collective sigh of relief, technology continued to advance by leaps and bounds, and particles were being chucked with abandon into the accelerator, inevitably leading to the discovery of many more subatomic particles – close to 200 in all. No one could see the wood for the trees any more. It was a true 'particle zoo'. How could anyone possibly discern an underlying system behind it? How could these particles, so wildly diverse, fit into any coherent framework? The carefully established order of particle physics was completely upended by a small bunch of overenthusiastic explorers. How dare they?!

Enter Murray Gell-Mann (1929–2019), a polyglot who spoke twenty-five languages, a walking encyclopaedia and an avid lover of nature. There wasn't a bird or flower he couldn't identify. At just twenty-one, Gell-Mann applied to work with Oppenheimer, but since the latter could not offer him a job, he started working with Fermi. Tragically, Fermi died not long after, and Gell-Mann eventually ended up as a fellow professor with Feynman at Caltech (the California Institute of Technology). With characteristic energy, Gell-Mann set out to tame the particle zoo of subatomic particles. If we can't immediately perceive a system, let's assume one! What if the bewildering array of particles was constructed from building blocks even smaller than the now familiar electrons, protons and neutrons? What if those particles had an underlying symmetry? And what if group theory could help to structure everything? Gell-Mann's vision extended far beyond Yukawa's mesons, which had been identified as the glue holding protons and neutrons together. He realized that the strong force didn't only operate between the protons and neutrons, it also worked *within* them. Gell-Mann's disruptive insight revealed that these nucleons themselves consist of even smaller particles: the quarks.

QUARKS IN SCENTS AND COLOURS

The discovery of quarks was itself a kind of superposition, with two scientists arriving at the same discovery at the same time, in 1964. Except that George Zweig called his quarks 'aces', inspired by the four aces in a deck of cards, while Gell-Mann found his muse in literature, that is to say James Joyce, whose *Finnegans Wake* begins with the line: 'Three quarks for Muster Mark!' This inevitably prompts the question: do physicists read novels? Or, put more gently: what's the connection? Perhaps it's the fact that the word 'quark' (an old English word meaning to croak, onomatopoeic for the rasping caw of a crow, which Joyce intended to rhyme with Mark but Gell-Mann meant to sound like 'quart') has a certain vagueness and multi-layered mystery, much like those atomic particles that went undetected for so long?

All those hundreds of newly discovered particles could actually be explained by assuming that the symmetry group SU(3) represents the superpositions of the three types of quark. No need to go into detail here. The main thing is that the beauty of the theory left no doubt about the existence of quarks any more. Although you can't see them – quarks cannot exist individually – they bring order to chaos. And being so small, they fit easily into a nutshell: all particles, including Yukawa's mesons (later called pions) are made up of quarks. It was later discovered that quarks exist not in three but in six distinct 'flavours': up and down, charm and strange, top and bottom (or beauty). But we mainly use up and down, referring to the spin. A proton contains one down-quark and two up-quarks, while a neutron contains two down-quarks and one up-quark. The weak nuclear force, for all its 'weakness', has the power to change a quark's 'flavour', such as flipping a down-quark into an up-quark. And this is precisely how and why a neutron turns into a proton. If you capture this process in a Feynman diagram, it looks something like this:

Beta decay: a neutron decays via a W boson into a proton, an electron and a neutrino.

Hold on a second. Quarks are fermions. And fermions can never occupy the same state. Then how can a proton possibly have two up-quarks? By introducing an additional degree of freedom: colour. Red, blue and white. What? White?? That's not a primary colour! All right: red, blue and green. It doesn't matter that much anyway. In the end, quarks have nothing to do with colour; the colours are merely mathematical values. If each quark has a different colour, then a proton can perfectly accommodate two up-quarks without breaking the rules.

With the quarks bringing some order in the chaos, the next question was: how do these quarks interact with each other? To answer this question we need yet another set of boson particles: the gluons. And what grand underlying formula could unify all this beautiful information? Again, Gell-Mann's ideas are perfectly compatible with the gauge theories of Yang and Mills. In this system, gluons act as gauge bosons, mediating the interactions, while the quarks serve as matter fields. Given the 'colours' of the quarks, the theory was renamed quantum chromodynamics (QCD). The strong force was (almost) under control!

The final piece of the puzzle was understanding why it's impossible to pull quarks apart. The answer: when you try to separate them, the gluon field stretches like an unbreakable rubber band. The further you pull them apart, the stronger the force pulling them back together. This phenomenon, known as 'confinement', also explains why you'll never find a solitary quark. They always appear in pairs or triplets. In a dashing feat of calculation – achieved in parallel in 1973 by Gerard 't Hooft, David Gross and Frank Wilczek, and David Politzer (clearly, the time was ripe!) – it was demonstrated that the forces between quarks in chromodynamics have a fascinating dual nature: they are infinitely strong at long distances, but also infinitely weak at short distances. This property, nicknamed 'asymptotic freedom', made it possible to use perturbation theory to describe quarks, and render the whole theory workable for making predictions. Long live Feynman diagrams! The predictions QCD made about the behaviour of quarks at very high energies have since been experimentally verified, tested and ratified. Theory and experiment matched, and quantum chromodynamics had been proven to work, despite the fact that, in the low-energy regime (where the truly interesting many-particle physics emerges), predictions remain frustratingly illusory. The reason: they are too difficult. Even calculating something relatively 'simple', like the neutron's mass, required cutting-edge supercomputers and a dizzying series of calculations. All grist to the mill of the quantum computer!

The next milestone was to make quantum chromodynamics

compatible with the weak and electromagnetic forces. This headway was made possible thanks to the work of Steven Weinberg, Abdus Salam and Sheldon Glashow, who demonstrated that all these forces fit together nicely under the graceful arrangement of the group SU(3)xSU(2)xU(1).

THE STANDARD MODEL

The holy trinity had been identified and is composed of three forces: the strong force, binding everything together (gluons and quarks), the weak force, responsible for radioactive decay (quarks and leptons) and the electromagnetic force, applied to charged particles (photons). Together, these forces are synthesized in the standard model, which describes the incredibly close interaction that exists between them, gravity notably excluded. The standard model remains the most accurate model for understanding particles and forces in the subatomic world. No experiment has (yet) been able to expose any flaws or gaps in its completeness.

One of the predictions of the standard model was that, alongside the W and Z particles (of the weak force), there had to be another boson linked to the strong force. As already revealed, the answer lay in symmetry breaking. Fast forward thirty years (and €10 billion), and in 2012 the W and Z particle were joined – to huge fanfare – by the Higgs particle, the 'heavy-maker' that gives mass to other particles. Thanks to CERN, the weight of the Higgs particle was measured with extreme precision, something that had proven to be rather difficult to predict. This triumph came fifty years after the prediction by Philip W. Anderson, famed for his quip that 'theoretical physics is applied group theory'.

It was only fitting that the 2013 Nobel Prize went to the Belgian François Englert and the UK's Peter Higgs, the scientists who (jointly with the American Robert Brout, who died shortly before the Higgs particle was discovered) had brought Anderson's prophetic words to life.

Let's play devil's advocate for a moment. Wouldn't it have been interesting if the Higgs particle hadn't existed after all? That would mean there is something in nature that we don't understand. True. But it – really! – does exist. As you can see, there is absolutely no magic or a crystal ball involved in predictions like these. Or, as Stevin so beautifully put it: 'Wonder is no wonder.' Everything has an explanation. What *does* remain a mystery is why the popular press insists on calling the Higgs particle a 'God particle'.

THE THEORY OF EVERYTHING?

Gerard 't Hooft demonstrated that the Yang-Mills theories, including the standard model, are renormalizable. With that insight, he made a major contribution to the domestication of infinity. But here's the catch: the moment we include Einstein's gravity in the mix, the standard model loses its renormalizability. The main reason is that gravity uses gauge particles (the 'gravitons') with a greater spin. And such theories cannot be renormalized. Which brings us to the million-dollar question in theoretical physics: how can the four fundamental forces of nature (strong nuclear force, weak nuclear force, electromagnetic force and gravity) be reconciled? What would such a Field Theory of Everything look like? The Theory of Everything remains the cherished dream of just about every theoretical physicist, much like Newton's quest for an 'Alchemy of Everything'. But for the time being, it remains a holy grail, leaving us to content ourselves with a 'Theory of Maybe Everything'. Meanwhile, is it even realistic to think that nature could be described by a single coherent, basic formula? A theory that describes both the intricate properties of fundamental particles and the staggering interactions between stars? The closest we have come to such a comprehensive theory is string theory. And that's no surprise: string theory states that the four fundamental forces were once a single fundamental force – until the Big Bang blew everything apart – and that all the point particles in the universe are actually unimaginably tiny, invisible strings, vibrating according to a specific pattern. Depending on the nature of this vibrational pattern, a particle acquires a mass and a charge, dictating whether it manifests as a quark, an electron or another elementary particle.

A few decades ago, this gave rise to 'superstring theory', also called M-theory, with the M standing for 'membranes' – very extended spatial planes. Cynics, though, suggest that it could just as easily stand for 'murky'. Superstring theory says that we live in a universe of not four dimensions (three of space and one of time) but rather eleven: three dimensions of space, one dimension of time and another seven spatial dimensions. These higher spatial dimensions are said to be tightly 'rolled up' in a space, also known as Calabi–Yau space, that is not visible to us. Superstring theory also introduced the concept of 'supersymmetry', where fermions are assigned to a boson, and bosons to a fermion, granting every elementary particle a 'superpartner'. The point is that this quantum mechanical theory also implies gravity. In the end, that's what it was all about. As far-fetched as it sounds, superstring theory is more than just another bold attempt to understand the universe's inner workings. Despite decades of intense research, however, progress has been frustratingly slow, with no experimental evidence yet to validate this super-everything-and-anything hypotheses. But for now, it *is* one of the few options we have at present. Benefit of the doubt then, so be it! After all, atoms too were once seen as some sort of futuristic magic trick, until they turned out to underpin everything and consistently confirmed experiment after experiment.

7.3 We are all made of stars

Time to plant our feet firmly back on the ground. We've come to the clear conclusion that the number of different types of atoms on Earth – and across the universe – is very limited. But of course, everything that is around us must have originated somewhere. So, where is that mysterious little factory that produced all these atoms?

For answers, we turn again to George Gamow, who had become solely captivated by the stars and the big-bang theory of yet another George, the Belgian priest, astronomer, naturalist and mathematician George Lemaître. Billions of years ago, before the Big Bang, the universe is thought to have been as inconceivably small as a single atomic nucleus, one with the mass of the entire universe. Lemaître's audacious idea of a 'primordial atom' tickled Gamow's fascination. Since this chief moment united the very largest and the very smallest scale, we inevitably have to resort to quantum physics to understand what truly happened. The prevailing theory suggests that, one fateful day, a super-radioactive process caused that (immensely unstable) primordial atom to explode.

> ### ALPHA BETA GAMOW
>
> George Gamow, born in what is now Ukraine, was an exceptionally cheerful chap. This eternal joker couldn't resist adding a twist to the 1948 paper 'The Origin of Chemical Elements',

co-written with his PhD student Ralph Alpher. He sneakily included the name of his good friend, physicist Hans Bethe, to create the authors' list 'Alpher, Bethe, Gamow' – a playful nod to Rutherford's Greek alpha, beta and gamma particles. Gamow, sloppy when it came to sums, had a habit of sprinkling a bit of levity and nonsense into serious academic work, much to the dismay of the meticulous Alpher, who by the way wasn't amused by the wordplay. Since most of his colleagues had been rounded up for the Manhattan Project, Gamow happily pursued his own cosmic curiosity: the Big Bang. In their paper on the origin of the elements, Gamow and Alpher elaborated on a Carl Friedrich von Weizsäcker theory that explains why hydrogen and helium make up 98 per cent of the sun's mass, with only a tiny fraction of heavier elements.

During the process Gamow dubbed 'Big-Bang nucleosynthesis', all the atoms are thought to be created by nuclear reactions, starting from hydrogen atoms. First on the list: the birth of helium. When two hydrogen atoms come extremely close together – a feat requiring tonnes of energy (with associated scorching temperatures) – they can fuse to form a helium atom. This process, called nuclear fusion, releases a tremendous amount of energy, much more than was needed to bring the hydrogen atoms together. This excess energy is emitted in the form of electromagnetic rays: the very sunshine that lights up our faces.

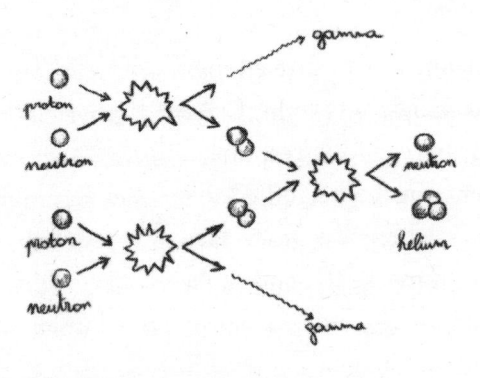

The birth of helium.

So how come the sun is still burning after all these billions of years? Because there's a staggering abundance of hydrogen in the sun. And because the chance of two hydrogen atoms fusing to form helium is very small (about one in a billion or so). To make this a bit more tangible: every second, the sun converts some 500 million tonnes of hydrogen into nearly as much helium.

This is all very formidable of course, and a huge step forward in our understanding of the universe, except it doesn't tell us how that remaining 1 per cent of heavier elements came into being. To know that, we turn to 1953, when Fred Hoyle (1915–2001) stormed into his friend Willy Fowler's office at Caltech and exclaimed: 'I exist! And I consist of carbon!' The surreal idea that all materials (and indeed all humans!) could have originated from a single star made him wonder what might happen if not two but three helium atoms (alpha particles) were to come very close together at exactly the same time and in exactly the same place. Hoyle calculated that those particles, like three tiny perfectly tuned violins, would then create a kind of unified vibration, converting them into carbon (C_{12}, more specifically). And then we're done, as carbon is a much heavier element. The chance of this process happening is shockingly small, but with an immense number of particles involved, the chances increase significantly. And sure enough, carbon is found everywhere in the universe. Hoyle was right. Using a similar approach, he identified the nuclear reactions responsible for the creation of the full array of elements in the periodic table – at least, up to number 26, iron (Fe). That's where the story takes a turn. When a star starts producing iron, it marks the end of its nuclear reaction. The reason: the energy it takes to create a new kind of atom far exceeds the energy that is generated.

This naturally leaves us wondering about the origins of the heavier atoms in the periodic table. The solution to this mystery comes with quite a bang. Once a star has fused enough iron, the outward electron pressure (Pauli's exclusion principle) can no longer counterbalance the inward gravitational pressure of the iron atoms, leading to an explanation packed with unavoidable superlatives. The core of the star implodes, and the outer regions are sucked inwards at dizzying speeds, driving the core's temperatures up to an astonishing 100 billion degrees Celsius. This extreme environment triggers the weak interaction, generating gargantuan numbers of neutrons and unleashing a freakishly strong outward force. This causes a titanic explosion: a supernova (a type II supernova, to be precise). In this cataclysm, neutrons are flung all around and crash into the iron nuclei, often in large clusters. The ensuing nuclear processes forge heavier elements, right up to uranium. Every atom beyond iron in Mendeleev's periodic table is born in the fiery heart of such a supernova.

But the story doesn't end there. We still hadn't figured out *why* there are so many heavy elements. That mystery wasn't cleared up until 2017. On 17 August of that year, two neutron stars collided. The observation was as surprising as it was indirect. But what was actually observed? Gamma rays and gravitational waves, released by the collision – and that observation was as inconceivable as it was unforeseen. Scientists had just witnessed a live collision of two neutron stars! Well, 'live' is of course a very relative term here, as the event actually occurred many millions of years ago, but the signals reached Earth on that precise day in 2017. It was a day worth its weight in gold. The observed spectrum revealed distinct peaks, leading scientists, with justified euphoria, to deduce that enormous quantities of gold, platinum and other precious metals had been created. In essence, the birth of the heavy metals had been observed that day. All the zinc in our bodies – and all the gold, silver and copper in the world – owes its existence to supernovas and colliding stars. But the opposite is also true: nuclear processes can convert heavy elements back into lighter ones. This process, called nuclear fission, occurs spontaneously, until the infamous number 26 (iron) is reached again. One day, far, far in the future, the universe might just become one colossal ball of iron.

To conclude: there are two types of natural nuclear reactions. Nuclear fusion converts lighter elements into heavier ones, while nuclear fission breaks heavier elements down into lighter ones. All current nuclear reactors operate on the basis of nuclear fission. Extensive research has been devoted to fusion reactors, because they promise far greater energy output and generate significantly less radioactive waste. Except that building a functional fusion reactor is a monumental challenge. The walls must withstand temperatures exceeding a million degrees, it's essentially like recreating the sun on Earth. Not only does that cost astronomical sums but, for the time being, it consumes far more energy than a reactor can give us back in return. This problem continues to provide scientists with plenty of food for thought. Meanwhile we can hold onto the profoundly unifying idea that we are all, quite literally, made of stardust. Every atom in our bodies was once created by a process with odds of no more than one in a billion, within the furnace of a distant star burning somewhere in the universe. Or in a gigantic supernova. Or in a collision of two neutron stars.

IN A NUTSHELL

> The behaviour of many particles cannot be explained through the study of single particles. New laws emerge at every scale. Death to reductionism!

> The renormalization group shows why physics is possible. Away with the details. Everything is symmetry!

> Symmetry breaking. Super conduction!

> Quantum Hall: perfection emerging from imperfection.

> *Dramatis personae*: Philip W. Anderson, Kenneth Wilson, Heike Kamerlingh Onnes, John Bardeen, Klaus von Klitzing.

EIGHT

MORE IS DIFFERENT

8.1 Emergence

It all started with what seemed like a pointless experiment at the home of Christiaan Huygens (1629–1695) in the Dutch city of The Hague. That day in 1665, feeling under the weather, Huygens crawled back into bed. He'd already binge-watched all the seasons of *Seinfeld* and there was no one to have a chat with on the phone, so he just lay there, staring straight ahead. He proudly admired the two pendulum clocks attached to a wooden plank in the adjoining room. His own handiwork. If one swung to the right, the other swung in unison to the left, perfectly out of sync yet harmoniously so. *Tick-tick tock-tock*.

Boredom can inspire mischief, and so, something suddenly prompted him to disturb that pendulum's ticking a little. To his amazement, he discovered that disrupting one clock caused the wooden plank to vibrate, disrupting the pendulum of the other clock. This wobbling continued until, as if by magic, both pendulums regained their balance and resumed ticking nicely in their faithful opposite synchronicity. Huygens, utterly fascinated, couldn't let the matter rest. He immediately devoted himself to a thorough study of this curious behaviour, determined to make sense of it. By now, one thing was certain: the whole is more than the sum of its parts. The interplay of the two clocks defied explanation. This could not possibly be explained by the behaviour of each clock individually.

We are now some 360 years further down the line. Huygens is feeling much better (thanks for asking) but to this day, nobody has succeeded in explaining exactly what was going on with those clocks. Which mechanism causes two clocks to get so thrown off? The moral of the story: just because you can explain the pendulum movement of one pendulum, does not mean you can understand, or predict, what happens when two pendulums are hanging side by side. Why is it so challenging to explain this synchronization problem – or call it cooperative phenomena – from a mathematical perspective? Because non-linear forces are in play here, and they require a completely new approach. The (very classical) phenomenon of non-linearity bears resemblances with the entanglements seen in quantum physics.

In the previous two chapters, we explored the reductionist philosophy in the quest for ever smaller particles. From molecules, we delved deeper into the atoms, where the old 'currant theory' gave way to a genuine atomic nucleus, housing a parade of protons, quarks and nuclear forces. By continually zooming in, from small to even smaller, reductionists like Einstein and Dirac ended up with the unification theory that could explain everything. Dirac concluded that quantum theory was now essentially complete, and that all the underlying laws required for physics (and chemistry) were known. The only problem was that the equations were too complex to solve. So, there you are. Armed with a whole bunch of formidable laws. That you cannot apply . . .

The Super Theory of Everything. It promised to be the ultimate magic formula, solving every mystery. The only problem: that theory does not yet exist, may never exist and – more importantly – that vision is fundamentally flawed. Reaching the quarks level doesn't mean we can suddenly explain the functioning of a superconductor. To quote Philip W. Anderson: 'The reductionist hypothesis does not by any means imply a "constructionist" one: The ability to reduce everything to simple fundamental laws does not imply the ability to start from those laws and reconstruct the universe.'

Reductionism may have led scientists to the fundamental laws, but science is not a two-way street. The idea that you can simply reverse course and use those fundamental laws to construct a full understanding of reality is both overly simplistic and naive. Yes, everything is connected, but somewhere between the very smallest and the very largest lies a boundary, something inherently irreconcilable. Like a Janus face, reality has dual identities that can hold conflicting properties. The wave–particle duality falls squarely into this same category.

No one understood this better than the aforementioned Philip W. Anderson (1923–2020), one of the most brilliant and influential physicists of the twentieth century. Anderson wasn't particularly concerned with the tiniest departments of quarks and string theories, although his work played a very important role in solving the Higgs boson puzzle. Nor was the largest scale, perforated with black holes, at the centre of his attention. What did interest him endlessly was that reign of the 'very many': very many atoms; very many electrons – in other words: many-particle systems.

Beyond the division between theorists and experimentalists, science can also be split into two additional categories. While some researchers delve ever deeper in search of ever smaller particles and fundamental laws, others pursue extensive research, seeking to explain phenomena that transcend those fundamental laws. High-energy physicists belong to the first category, while the second category includes condensed-matter physicists and biologists. Philip W. Anderson, in every fibre of his being, firmly belonged to the latter category.

Philip W. Anderson

In his seminal essay, 'More Is Different', Anderson argues that the organizing principles and laws we observe depend entirely on the scale of the system. This phenomenon is known as *emergence*. While it may be perfectly possible to understand a single atom or three quarks, predicting how a multitude of atoms or quarks will behave is a completely different challenge. The behaviour of the many cannot be reduced to the behaviour of the unity. The real question becomes: how do we deal with a system composed of countless particles? Instead of digging deeper and deeper into the nature of one particle, we must shift our focus to systems of *many*. And by many, we mean very many: 10^{24}, say.

EMERGENCE

A fish can swim swiftly, but a school of fish moves with a totally different dynamic. A lone starling flits about, but a murmuration is almost magical, and no one can pinpoint the bird that first decides to change direction, prompting the rest to instinctively follow. One person is only a person, but a community creates a city. And it's impossible to predict how a city will develop. Sometimes it just takes one cafe in a run-down neighbourhood, and in no time it attracts a theatre, then a bike repair shop, followed by the inevitable coffee shop. And just like that, the street is hip!

Turn to the nearest piano and strike a note, say, B-flat. It might sound fantastic enough, but it's unlikely that it will stir any great emotions. Now hear that same note in Beethoven's *Hammerklavier* sonata, and we hope the hairs on your arm rise in collective awe. The point is, mastering a single B-flat does not mean you can automatically conjure Beethoven's *Hammerklavier* or, heaven forbid, his second Piano Concerto. A piano tuner works note by note, but in sonatas and symphonies, it's not about individual notes. What truly matters is the interplay between notes and musicians, the rhythm, the dynamics, the emotion and the interpretation of that emotion. The music, *quoi*.

In essence, it always comes down to asking the right questions. Why are certain materials superconductive? What role does chaos play in a substance? How can you construct a mathematical formula that is both simple enough to capture the essence, yet complex enough to describe the universal, organizing properties of a many-particle system? And how do you do so in such a way that these universal properties do not depend on the microscopic properties of the system, but transcend them? After all, it does seem possible to describe very different systems using one and the same formula, provided they share the same symmetry. Take, for example, how the magnetic field in a material changes with the temperature. It corresponds to the same fundamental formulas that describe how water turns into ice. And so, once again, we end up back at symmetry breaking. Every phase transition involves either the breaking or restoration of symmetry. Symmetry breaking is a cooperative – that is to say, emergent – phenomenon. It arises only in systems composed of very many (interacting) particles. Therefore, it cannot be explained through the study of individual particles. When we zoom out to look at the properties on an increasingly macroscopic scale, we see that systems behave differently, depending on the scale. This is the hallmark of emergence. Think of the difference between biology and particle physics: you (thankfully) don't need gauge theories to understand biology.

Each level or scale therefore operates according to its own fundamental laws, which are only indirectly influenced by the particularities of the underlying step(s). Of course, while the steps are consistent with respect to each other, the properties of a higher level cannot be directly derived from those of a lower one. Compare it to a skyscraper where each floor has its own unique laws. The elevator connects all the floors, but the transitions between them remain unpredictable. The only thing that all the floors have in common is the shell, the foundations.

Physics follows a similar principle. We can explain a broad spectrum of phenomena using a very limited number of formulas, because the 'glue' binding the different levels consist of symmetries or groups, and these are inherently limited. This explains why the physics of entirely different systems can still be described using the same groups. All phase transitions can essentially be reduced to one of the limited possible forms of symmetry breaking.

We must, therefore, rethink the meaning of the term 'hierarchy'. It's all too easy to say that one branch of science naturally stems from another, or that one is more important than another. Knowing the fundamental laws doesn't mean that you can automatically understand and predict all natural phenomena, whether at the micro or macro level. One isn't superior to the other, it is simply different. It's easy for a physicist to crack open a chemistry book and bookishly think they understand it all. The reverse is already somewhat trickier. And while chemistry might seem, at first glance, nothing more than applied quantum physics, a physicist is unlikely to understand it. Their physics intuition doesn't tell them which chemical processes will leap out of Pandora's chemical box. Psychology isn't just applied biology, and biology isn't merely applied chemistry. Each level demands its own approach and brilliant insights to understand which laws apply there. Yet mathematics – and especially the symmetries – remain universal across disciplines.

The principle of universality fuels ongoing tensions between high-energy physicists and others in the field. Anderson's contributions were crucial in this regard, as he initiated a real discussion – not only among scientists, but also among philosophers – weighing the merits of digging deeper versus casting wider nets. A perfect illustration of this tension is the debate between Anderson and Steven Weinberg. Weinberg, maestro of the smallest particles and architect of the standard model, was convinced that, at least in theory, the macroscopic world could be explained on the basis of microscopic laws. The discussion took place in 1993. The United States was to build its own particle accelerator, the Superconducting Super Collider (SSC). Cost of this megalomaniac tunnel: $5 billion. The plan had its supporters and opponents, even within the various echelons of the scientific world itself. Location of the debate: the US Congress. Anderson argued that far too much money was being poured into elementary particle physics and far too little into many-particle physics. Things ought to be a bit more democratic in the supermarket of sciences. The discussion captures the essence of the discourse with striking clarity. Judge for yourself.

WEINBERG: *I am grateful to the chairman to allow me to come here to talk about the Super Collider. In essence, the Super Collider is a machine for creating new kinds of matter, particles that have existed since the Universe was about a trillionth of a second old. To produce these particles requires an energy about twenty times higher than the energy of the largest accelerators that now exist, which is why the Super Collider is so big and therefore why it is so expensive. This little statement that I have made really does not do justice, however, to what the Super Collider is about because particles in themselves are not really that interesting . . . If you have seen one proton you have seen them all. We are not really after the particles, we are after the principles that govern matter and energy and force, and everything in the Universe. Culminating around the mid-1970s, we developed a theory called the standard model which encompasses all the forces we know about, all the different kinds of matter that we can observe with existing laboratories. We know that [this theory] is not the last word [because] it leaves out things that are pretty important, like the force of gravity . . . In addition, the particles that we know, quarks, electrons and so on all have mass . . . But [the theory] does not know exactly what [these masses] are. This is the question that the Super Collider is specifically designed to answer. But there is a sense, nevertheless, [that] this kind of elementary particle physics is at the most fundamental level of science. That is, you may ask any question, for example, how does a superconductor work, and you get an answer. You get an answer in terms of the properties of electrons and the electromagnetic field and other things. And then you ask, well, why are those things* *true? And you get an answer in terms of the standard model . . . And then you say, well, why is the standard model true? And you do not get an answer. We do not know. We are at the frontier. We have pushed the chain of why questions as far as we can, and as far as we can tell we cannot make any progress without the Super Collider. Thank you.*

CHAIRMAN: *Thank you very much, Dr Weinberg. Our next witness is Professor Philip Anderson from the Department of Physics, I think that is Applied Physics, at Princeton University.*

ANDERSON: *I will try to be as brief as possible and in any case I do not think I can be anywhere near as eloquent as my colleague here, Steve Weinberg.*

WEINBERG: *You can try.*

ANDERSON: *The point of my testimony is priorities. The physics being done by the SSC is in a very narrow specialized area of physics with a very narrow focus. It focuses on the very tiny and very energetic sub-sub-substructure of the world in which we live. Most of that substructure is well understood in a very definite sense. Nothing discovered by the SSC can, for the foreseeable future, change the way we work or think about the world and cannot change even nuclear physics. Perhaps a couple of hundred theorists (too many for such a narrow subject in my opinion) . . . and a few thousand experimentalists work in this particular field of science. That is less than 10 per cent of the research physicists in the world . . . Yet the budget of [the SSC] dwarfs the budget for all the rest of physics. The fact is that particle physicists are funded, on average, ten times as liberally as other physicists . . . In this* *sense, the SSC is not a very efficient jobs program, at least for physicists. At least two books and many articles have been published recently trying to justify the special status for this particular branch of physics as somehow more fundamental than all other science. That so many particle physicists have time to write such books and articles may tell you something about the real interest in the field; it has not made much progress lately, and so they do not have anything else to do. There are many other really exciting fundamental questions which science can hope to answer and which people like myself are, on the whole, too busy to write books about. There are questions like: How did life begin? What is the origin of the human race? How does the brain work? What is the theory of the immune system? Is there a science of economics? All these things have in common that they are manifestations not of the simplest things about matter – the elementary particles – but of the complexity of matter and energy as we ordinarily run into them. These manifestations of complexity do not . . . have any possibility of being affected by whatever the SSC may discover . . . On the other hand, the future seems to me to belong to these subjects, to these questions, rather than to the infinite regression of following the tiny substructure of matter. Perhaps you should think which fundamental questions are easier and less expensive to solve. Thank you.*

Two months later, the US Congress cancelled the SSC.

8.2 Renormalizing

In Chapter 7, we explored renormalization theories in depth. What began as a mathematical trick to tame infinity, paving the way for field theory's many successes, turned out to have much more to offer. One day, Kenneth G. Wilson (1936–2013) unearthed this particular insight, an epiphany so exhilarating that it inspired him to break into a little folk dance (a pastime he greatly enjoyed).

We've already argued that every scale operates under its own set of laws (and complexities), with unique organizing principles emerging at every level. Wilson, through his pioneering use of the renormalization group, and his method of rescaling space and time (essentially zooming out), explored how one theory, or Hamiltonian, can transform into another. To recap: a Hamiltonian describes the kinetic and potential energy of particles and their interactions. Wilson differentiated between three possible types of terms in the Hamiltonian: the relevant, the marginal and the irrelevant. Relevant terms grow in importance as the scale increases, irrelevant terms diminish and the marginal terms remain somewhere in the middle. For example, the energy scales required to describe what happens inside the nucleus (the degrees of freedom of the individual protons and neutrons) are totally irrelevant when it comes to describing the behaviour of a molecule.

On a large scale – because that is ultimately what interests us – the irrelevant terms become negligible, while the relevant terms define the possible degrees of freedom. In the end, all of physics boils down to the marginal terms. From this, Wilson deduced that the ultimate microscopic theory, the holy grail for every reductionist, is actually completely irrelevant, since it consists almost entirely of irrelevant terms. At large distance scales (meaning 'large' relative to the distance between separate particles), many-particle physics is therefore determined entirely by a very limited set of possible marginal terms.

Although each scale generates its own theory, when we look at the interaction between those various scales and the nature of phase transitions, there is only one thing that connects them, something no theory can escape: the symmetries inherent in the system. This brings us full circle to the claim we so boldly trumpeted in Chapter 2: symmetries are the most powerful organizing principle in physics. They bridge the gap between the various scales and are crucial to understanding phase transitions. Emmy Noether and Lev Landau were quite right: physics is essentially the study of symmetry.

No matter how groundbreaking the discoveries at the very smallest scale may be, to truly understand how nature works as a whole, we must focus on organizing principles that persist across all scales and are independent of the 'details' from microphysics. We already cited the example of magnets and water/ice: completely different properties (one pertains to spins, the other to molecules), yet their phase transition is described in exactly the same way. Why? Because they share the same symmetry. Symmetry is what endures. Symmetry is what remains. Symmetries are for ever.

This leads us to the astonishing conclusion that, in the end, we need only a very limited number of fundamental, simplified, universal laws that can be used to explain virtually everything. And yet, paradoxically, reductionism tells us nothing about this. The physicist Leo Kadanoff (1937–2015) confirmed this point: 'All the richness of structure observed in the natural world is not a consequence of the complexity of physical law, but instead arises from the many-times repeated application of quite simple laws.'

In conclusion, although the theory was initially dismissed as yet another counterintuitive and 'ugly' (to borrow Dirac's own word) intermediate solution from physics' box of tricks, as a 'weapon' to fend off unruly infiniteness, we are left with no choice but to acknowledge that it is thanks to renormalization that physics is even possible. The exact nature of the underlying microscopic laws doesn't matter, because at larger scales only a limited number of possible theories remain, and these are fully defined by the symmetries they exhibit.

In this sense, Wilson's legacy is priceless. With his renormalization, physics (once again) entered a new era, an era where the perfection of macroscopic systems with countless particles became not just observable but understandable. This perfection arises because the emergent, collective properties of such systems can be captured by a wave function with flawless symmetry. And symmetry, as it turns out, is the real star of the show in the renormalization group: it's either present or it's not, and nothing else truly matters. Two wonderful examples of this unpretentious perfection are superconductivity and the quantum Hall effect.

8.3 Superconductivity

Thanks to quantum physics, it became possible to explain the existence of atoms and, by extension, also the structural properties of matter. However, it's a misconception to think that quantum effects occur only at the microscopic level, invisible to the naked eye. At the macroscopic level too, these effects can sometimes manifest. The most striking examples are superfluidity and superconductivity.

In the early 1940s, physicists were fascinated by superfluids. Superfluidity was discovered during experiments with helium, which under certain conditions exhibits quite a few Bose–Einstein condensate allures. Let's run a little experiment. Take a glass, half-fill it with helium-4 (bosons) and slowly lower the temperature. At some point, a phase transition takes place. Wait a moment, and then witness the helium suddenly becoming a superfluid, slipping stealthily away over the edge of the glass like a thief in the night. The superfluid no longer has any viscosity, and since the helium molecules are attracted more by the edge of the glass than by each other, they begin to creep upwards. How can this be explained? What happens during such a transition? These are exactly the questions that intrigued Lev Landau. This curiosity led him to his trailblazing theory of symmetry breaking, a concept that extends, of course, far beyond this specific problem. In the case of a superfluid, the symmetry breaking involves a gauge field, and its mathematical description is exactly the same as that of a Bose–Einstein condensate – and of a superconductor.

The superconductor is a remarkable example of a quantum many-particle system with extraordinary properties. Superconductivity is an emergent phenomenon, meaning it requires a different type of physics to explain it, since it cannot be deduced directly from microscopic theory (each scale obeying its own laws). This also explains why it took so long to truly understand the phenomenon. A superconductor is a material that doesn't offer any electrical resistance whatsoever below a certain temperature. The first person to observe this phenomenon was the Dutch physicist Heike Kamerlingh Onnes (1853–1926). It happened in 1911, the year of the first Solvay Conference, which Onnes himself also attended (he is standing to Einstein's right in one of the iconic group photos). A few years earlier, he had succeeded in liquefying helium gas. 'Breaking news!' he thought. But no, his theory full of superlatives didn't really resonate in Brussels at the time. That legendary conference raised more questions than it answered. Much of what was proposed only acquired its full significance years later. It would take decades to realize that a superconductor is much more than a perfect highway for electricity.

When Onnes cooled mercury, its electrical resistance – as expected – decreased. But the next day, when he lowered the temperature a little further, the resistance disappeared completely. It turned out that mercury becomes superconducting at temperatures below 4.2 Kelvin (just shy of −269 °C). And then, in a flash, Onnes saw Kekulé's snake winding through his mind. In a closed system made of mercury, current should flow indefinitely, without external input and, above all, without loss of energy. That *was* super!

Of course, other heavy metals also exhibit superconducting talent, but their transitions occur at different toe-curlingly low temperatures, known as the 'critical' temperature (or jump point). For aluminium, this jump point is 1.20 Kelvin (−271.95 °C), lead makes the transition at 7.19 Kelvin (−265.96 °C) and zinc at 0.87 Kelvin (−272.28 °C). Gold, silver and copper don't fall under the superconducting spell. Neither do nickel, iron and cobalt. These materials don't have any special magical powers.

So how does a superconducting process actually come about? In metals, atoms are arranged in a crystal lattice. Electrons from the outer energy orbitals of an atom roam freely, continuously jumping from one atom to another. When an electric current flows through a metal, these electrons must make their way through that lattice. But with the countless obstacles in their path, like vibrating atoms, this journey is anything but smooth. How are they supposed to shoot unhindered towards the positive pole if they are constantly deflected in the wrong direction? All this friction, clashing and colliding results in a lot of energy loss, routinely released in the form of heat (electrical resistance). In the case of a superconductor, however, there is no friction. None at all. Here, conduction is perfect. No energy is wasted. The answer to how such a thing was even possible came about fifty years after Onnes, in 1957, with the development of the BCS theory, named after its creators' initials: John Bardeen, Leon Cooper and Robert Schrieffer. This trio imagined the electrons to form pairs: Cooper pairs. Of course, that pairing gave them entirely new properties. Only, how can two negatively charged electrons possibly attract one another?

A crystal lattice is full of ions: electrically charged atoms lacking electrons (since their outer shell electrons have gone for a walk). When a negative electron passes a positive ion in that lattice, it generates a vibration in the form of a phonon (a quick nod to

Einstein, who predicted phonons back in 1907), creating a ripple effect that allows another electron to settle in its wake. In essence, two electrons can attract each other and form a pair, via the exchange of phonon particles. This is what makes a Cooper pair. Phonons, these tiny wave packets created when an atomic lattice starts vibrating, are the indispensable intermediaries – the positively charged glue – allowing two electrons to transform into a single bosonic particle.

Compared to a single electron, Cooper pairs have an extremely large de Broglie wavelength, allowing them to effortlessly bypass all obstacles. But they also have the critical property of being bosons, which allows them to condense into their Bose–Einstein ground state, provided the temperature is low enough (symmetry breaking!). On a macroscopic scale, this leads to a remarkable phenomenon: a perfect, perpetual current that, once initiated, can't be stopped. This rigidity is a hallmark of symmetry breaking, much like the hardness of a crystal, also the result of symmetry breaking. This supercurrent is 'carried' entirely by the Bose–Einstein condensate and cannot be explained in terms of individual particles (or even pairs of particles) navigating through a lattice. In truth, Cooper pairs don't move at all. What changes over time is the *probability* of encountering a Cooper pair at a given location. They all exist in the ground state, which is strongly influenced by the boundary conditions – in this case: the amount of magnetic flux (energy) passing through the loop. As a result, no energy is lost. It is like an electron staying in its 1S orbital around a hydrogen atom nucleus, albeit on a macroscopic scale. And this phenomenon is, all things considered, quintessentially emergent.

A fascinating consequence of symmetry breaking in superconductors is the expulsion of magnetic fields. Within the material itself, there is simply no room for them, as the photons from the electromagnetic field gain mass as a result of the symmetry breaking. The ability of Bardeen and co.'s theory to precisely explain this so-called Meissner effect played a major role in convincing the ever-sceptical physicists to embrace their model.

The magnetic properties of superconductors have numerous applications, such as the Maglev train (derived from – seriously now – magnetic levitation). Or MRI scanners. This type of scanner requires fabulously powerful magnetic fields that can only be generated by sending a very strong current through the doughnut-shaped superconductor. The same principle applies to the particle accelerator at CERN, where superconductors generate the magnetic fields that keep particles circling at vertiginous speeds, until they collide head-on.

JOSEPHSON JUNCTIONS

Superconductivity is also at the heart of the 'Josephson junction' (named after Brian Josephson (b. 1940)), one of the workhorses of modern quantum technology. This junction consists of two layers of superconducting material separated by a very thin layer of insulator. Where we might expect particles to collide with the insulator and bounce back, the Cooper pairs do not encounter any resistance. Thanks to quantum tunnelling, they pass through with ease. If the (constant) voltage U applied between the two superconducting materials exceeds a critical value, an oscillating current emerges with a frequency precisely equal to $2.U.e/h$ (e being the charge of an electron and h Planck's constant). This means that the ratio e/h, involving these elementary constants, suddenly becomes

> measurable on a macroscopic level. A phenomenon can't get any more emergent than that! What's more, this experiment is fantastically reproducible because it is completely independent of the specific properties of the material – yet another manifest property of emergence. Josephson junctions have many applications. They serve as the international reference standard for the volt. They are used in very precise magnetometers. And in single-electron transistors. And, of course, in quantum computers.

As we've seen, superconductivity typically requires extremely low temperatures. However, high-temperature superconductors have since entered the scene. But of course, 'high temperature' is a very relative term, because at 90 Kelvin (about −183°C), we are still well below freezing. The advantage of this type of superconductor is that it does not necessarily have to be cooled by liquid helium. Liquid nitrogen, which is considerably cheaper, can do the job just as well. Ideally, of course, the real breakthrough would be a superconductor that works at room temperature. Mobile phones and computers would consume virtually no energy, since electrical energy would no longer be converted into heat. Unsurprisingly, a lot of money and, well, energy, is therefore being invested in research into superconductors and green technologies. The problem is that the mechanism behind high-temperature superconductors is very complex and still not fully understood: these quantum states cannot be described using traditional methods like Hartree–Fock models and Feynman diagrams. And this is one of the reasons why the quantum computer is so eagerly awaited; it could play a paramount role in the development of better superconductors and unlocking their mysteries.

8.4 The discovery of perfection

In 1879, Edwin Herbert Hall (1855–1938) conducted an experiment – eighteen years, mind you, before the discovery of the electron (in 1897). Hall placed a thin rectangular gold leaf perpendicular to a very strong magnetic field. In the illustration below, we imagine this magnetic field to be directed in the B direction. He attached battery clamps along the A direction, allowing the current to flow through the strip. This setup spontaneously generated a weak voltage in the C direction. This phenomenon, known as the Hall effect, refers precisely to this potential difference. However, the voltage was so minuscule that it didn't stir much excitement. Scientists didn't see its full potential.

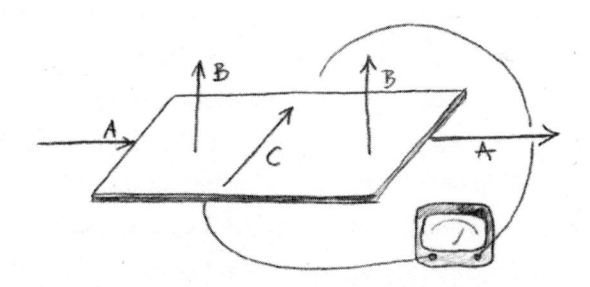

The Hall effect: a current flowing in the A direction, combined with a magnetic flux in the B direction, spontaneously generates a voltage in the C direction.

In the second half of the twentieth century, that subtle magnetic hiccup (finally) became a serious subject of further research, propelled by the alliteratively named Klaus von Klitzing (b. 1943). On the night of 4 to 5 February 1980, at around 2 a.m., von Klitzing discovered something that was never meant to be discovered. Something that no theorist thought possible. Von Klitzing discovered . . . perfection.

A quick recap. Von Klitzing, familiar with the Hall effect, was intrigued but not satisfied. His curiosity pushed him further. What if, he wondered, I perform this experiment at an extremely low temperature? What would happen to the electrons at the edge of that material? After all, the edge often exhibits different physics from the 'bulk' beneath it. As expected, the interplay between the current and the magnetic field spontaneously created a voltage difference in the C direction. But the results defied classical physics. The curve that he observed, representing the difference in potential, did not rise smoothly and linearly as a function of the magnetic field. Instead, it climbed in steps, moving through discrete levels. That's 100 per cent quantum! But that's not all: this quantized Hall current cannot be understood in terms of individual particles. It is the collective quantum ground state that 'glides' across the lattice.

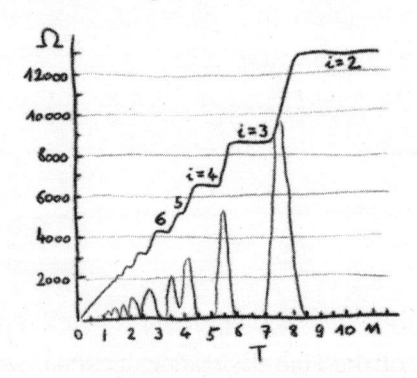

Each step represents an energy level. At each step, an electron band is completely filled. As the voltage continues to increase, band after band is filled.

The crux of the essence is this: no matter what material is used to perform the experiment, the result (the curve) remains exactly the same. This is the perfection of von Klitzing. The caveat, of course, is that your sample must be large enough and, strange but true, contain small amounts of impurities. It's precisely these impurities that allow us to see the quantum Hall effect. The 'imperfections' are not an obstacle, but a prerequisite for success, however counterintuitive that may seem. This is the power of emergent behaviour (*more is different*). Emergence leads to the most unexpected phenomena. Here we have perfection born out of imperfection; a Greek tragedy rewritten as a triumph!

MIND THE GAP

Powerful magnetic fields pull the electron bands of a material apart, creating sizeable 'gaps' between them. Impurities introduce new, localized orbitals where electrons can become trapped, a phenomenon known as 'Anderson localization'. The energy of these electron orbitals is contained within those very gaps. This enables the voltage in the system to be continually tuned, without it having to jump from one energy band to another. And this continuity is essential for the observation of the plateaus.

But what exactly did von Klitzing measure? His experiment revealed that the plateaus of the Hall resistance are quantized, and that the measured resistance value is always a fraction of the fundamental value $\frac{h}{e^2}$ = 25812.807 Ohm (Ohm being the unit of electrical resistance). The plateaus occur at the values $\frac{h}{e^2}$, $\frac{h}{2.e^2}$, $\frac{h}{3.e^2}$, $\frac{h}{4.e^2}$ and so on. This formula contains two fundamental constants: h (Planck's constant) and e (the charge of an electron). And that, as in the case of Josephson junctions, is actually quite striking: an experiment carried out on a macroscopically large system allows us to determine the individual charge of a single electron with extreme precision and robustness – provided, of course, that we already know Planck's constant. But the reverse is equally true: the experiment allows us to define Planck's constant in terms of the electron charge and the quantum Hall resistance. Namely, as e^2 times the quantum Hall resistance (25812.807 Ohm). Furthermore, the quantum Hall effect is used to calibrate electrical resistances. This discovery didn't require a complicated experiment or anything like a particle accelerator. Von Klitzing simply looked at the collective (emergent) behaviour of a vast number of particles to determine, with touching precision, the properties (especially the charge) of a single elementary particle. Emergence doesn't get more impressive than that.

A final note on perfection – now that we have finally found it, let's not let it slip through our fingers. There is a mathematical explanation for the fact that the quantum Hall effect is so perfectly reproducible and so robust. This robustness stems from the topological properties of the electron orbitals present in the system. Topology, a branch of mathematics, deals with properties that remain unchanged when a system undergoes slight transformations. Depending on their symmetry (what else?), a natural number known as the Chern number can be assigned to the collection of electron orbitals in a band. This number entirely determines the quantum Hall resistance. Being a natural number, it is inherently robust: it cannot transform continuously from one value to another. This is the essence of topology. In the quantum Hall effect, the electron orbitals may be distorted, but as long as the symmetry/topology of the orbitals remains intact, the result (the quantum Hall resistance) doesn't change either. Why not? Because the Chern number is maintained, preserving the precision and robustness of the phenomenon.

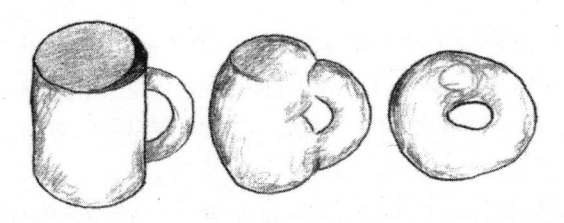

Topology of the coffee cup: without breaking its original structure, a coffee cup can be reshaped into a doughnut purely by maintaining its topologically invariant properties. You can stretch, fold and bend as much as you like; as long as you don't break or shatter it, anything goes. This beautifully reinforces another defining property of emergent systems: the details are entirely irrelevant.

The quantum Hall effect has an unexpected sequel. Von Klitzing cursed under his breath, frustrated that it was not he, but Horst Störmer (b. 1949) and Daniel Tsui (b. 1939) who had made this follow-up discovery, with Robert B. Laughlin (b. 1950) providing the theoretical explanation. Perhaps it was some consolation that the discovery was purely serendipitous, but even more than chance, it was the choice of materials (plus perhaps an advanced sense of curiosity) that proved decisive. In Störmer and Tsui's experiment, they decreased the temperature even further and used more powerful magnetic fields. Suddenly, and to their great surprise, the curve began to behave oddly, revealing levels *between* the steps. These intermediate steps led to the discovery of the fractional charge. The only plausible interpretation was that under specific conditions, particles or electrons split into quasiparticles. It turns out that even those fundamental, seemingly inseparable particles are capable of family expansion . . . These new parts-of-particles are neither fermions nor bosons. Frank Wilczek called them anyons (from the word 'anyone'). Split a particle into three, and you get three anyons, each with a charge of one-third. Although these anyons can be far apart, they always remain mysteriously linked by an invisible, highly elastic 'thread'. Elementary particles in the standard model do not exhibit such properties – anyons are truly emergent particles, and their very existence is only possible because of the many-body entanglements and correlations.

For fermions, wave functions get a minus sign when the particles are permuted (their order is swapped). However, this is child's play compared to what happens with anyons; when two anyons are permuted, their wave function transforms into a superposition of multiple states. This transformation occurs because the invisible threads connecting the anyons become entangled, resulting in exponential complexity.

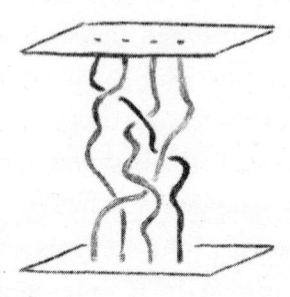

A braid of anyons. By twisting anyons around each other, you create entanglements.

That's all incredibly impressive. But what practical use can we make of these anyons? As we will see in the next chapter, one of the biggest technological challenges of the twenty-first century is building a quantum computer. The first major hurdle is that quantum states, especially superpositions, are extremely hard to isolate from unwanted environmental interactions. Designing a computer that operates on anyons and their permutations, could, in principle, overcome this obstacle, since the superpositions would then be protected by the intrinsic topological properties of the anyons. In practice, however, detecting and manipulating anyons in a lab has proven to be extremely challenging. Still, it does illustrate how much remains undiscovered, and how limited our knowledge of the quantum world of many-particle systems and their entanglements truly is.

IN A NUTSHELL

> Quantum metrology: quantum TikTok.

> Quantum simulators: a universal lab with 1,001 buttons to press.

> Quantum information and quantum cryptography: the monogamy of entanglement.

> Quantum computing: Shor's parallel quantum complexes.

> Quantum computers: welcome to the quantumdome.

> Quantum error correction: anyons, anyone?

> Quantum rethought: in information, I am.

> Quantum many-particle systems, revisited: tensor networks and quantum gravity.

> *Dramatis personae*: Richard Feynman, Peter Shor, Ignacio Cirac, Peter Zoller, Alexei Kitaev, Charles H. Bennett, John Preskill and the Quantum Computer.

NINE

THE SECOND QUANTUM REVOLUTION

On 5 June 1995, the second quantum revolution officially began. That day, the first Bose–Einstein condensate was created in a lab in Boulder, Colorado. Less than a month later, in that same year, in that same city, the first prototype quantum computer was built, featuring . . . two qubits. A new era had dawned. Suddenly, the whole world was under the spell of quantum.

Schrödinger and co. could never have imagined a future where individual atoms would not only become visible but manipulated – let alone that we could entangle them and use them to create entire new technologies. Seventy years after the advent of quantum physics and the prediction of the Bose–Einstein condensate, the world was ready for a second quantum revolution. Because now, it was undeniable: quantum could – no, had to! – become an even closer ally. After all, entanglements open up a new world of possibilities, provided we rethink the computer and everything in and around it. Quantum systems hold immense potential for processing information far more efficiently. But how? That is the question driving a whole army of quantum revolutionaries. Their mission: to rethink quantum physics. Armed with questions like: what does it truly mean to measure something? Where does the classical world end and the quantum begin? What is a computer anyway? And to say that something is complex – what does that even mean any more, in an era when so much human brainwork is left to algorithms? What language (vocabulary + grammar + semantics) can we use to describe entangled particles?

9.1 Quantum metrology

Historically, metrology – the science of measurement – has long been the bedrock of physics. The first clocks enabled drop experiments, which in turn led to Newton's laws. Measuring the speed of light culminated in the theory of relativity. Precise measurements of black-body spectra gave rise to quantum physics.

On the one hand, the precision of any measurement is defined, and therefore restrained, by quantum physics; Heisenberg's uncertainty principle sets the ultimate limit on measurement accuracy. It doesn't get any more accurate than that. On the other hand, the quantization of energy levels and the indistinguishability of particles is a blessing for reproducibility. Every measuring instrument made from the same atoms delivers identical results. The highest-precision sensors for measuring time, forces and electromagnetic fields are all based on this quantization. In the previous chapter we explored a number of cutting-edge electrical and magnetic sensors based on superconductivity (the Josephson effect) and the quantum Hall effect. We continue with yet another ticking example: the atomic clock.

Atomic clocks are quantum, through and through. They are tuned to the cadence of the impeccably stable and unperturbable frequencies of atoms – frequencies determined by the energy levels of the atomic orbitals. On closer inspection, atomic clocks work on the same principles as Huygens' pendulum clocks. Except now the pendulums are electrons. Atomic clocks tick with such unparalleled precision that they would only need to be adjusted by a single second once in every 100 million (10^8) years. This is why the SI unit of the second is based on the atomic clock. The world's most accurate atomic clock (in Boulder, Colorado) achieves an astonishing precision of 10^{-18} seconds. This clock operates on strontium atoms and is so sensitive that even a height difference of barely a centimetre has a marked effect on the clock frequency, a direct consequence of Einstein's general theory of relativity. If satellites weren't equipped with atomic clocks, GPS systems simply wouldn't work: the clocks in those satellites must measure the time difference between two signals sent just a few metres apart on Earth, demanding an extraordinary level of precision.

SYNCHING TO THE RHYTHM OF THE ATOMIC CLOCK

The SI unit of the second is defined as 9,192,631,770 periods of the 'pendulum' of the caesium atomic clock. The energy difference between the ground state and the first excited state of the caesium atom is precisely $\Delta E = h \cdot 9{,}192{,}631{,}770/s$. This frequency falls within the microwave range of the spectrum (similar to the frequencies in microwave ovens). To build a caesium atomic clock, caesium atoms are captured and held in a cavity (with the challenge being to keep them stable) where they are exposed to light generated by a microwave generator with tuneable frequencies. If the frequency of the microwaves matches the resonance frequency of the caesium

> atoms, the atoms will absorb a great deal of energy. If not, they won't. The frequency of the microwaves is therefore tuned to the specific resonance at which the atoms emit the most energy. The number of oscillations of the microwaves is counted by a separate electronic circuit, ensuring the clock's extreme accuracy.

9.2 Quantum simulation

With the Bose–Einstein condensate, it became possible to peer into the soul of matter for the very first time. We now understand pretty much everything about the internal workings of a particle – and yes, we do mean a single particle – but throw a whole bunch of them together, and no one can predict with certainty how they will behave. Emergent remains emergent. What's more: no one has yet proven Dirac wrong when he said, 'Just because we know the equations doesn't mean we can work with them.' Equations that are too complex may look very sexy, but if we can't use them, they're useless.

The aim of quantum simulation is quite simple: to solve quantum problems that neither the human brain nor even the most powerful classical computer can handle. Someone who wholeheartedly embraced this challenge was the famed and fearless Richard Feynman. As early as 1981, during a lecture organized by IBM and MIT, he launched a spirited, if not prophetic tirade, declared in his *vewy Amewican* accent: 'Nature isn't classical, dammit, and if you want to make a simulation of Nature, you'd better make it quantum mechanical, and by golly it's a wonderful problem, because it doesn't look so easy.' Which is tantamount to saying: do what you like, but if you want to simulate nature, you'll have to do it with quantum, because nature *is* quantum. And *yippie yo yippie yay*, what a wonderful problem! The harder it is, the more fun – and the more interesting!

In short: if bits won't do it, try qubits. Feynman was the first to take a constructive approach to these pesky quantum effects emerging in the ever-shrinking computer transistors. While most saw them as troublesome glitches (after all, who wants a smartphone that lets you swipe left and right at the same time?), Feynman saw things differently. What you can't avoid, you must embrace.

THERE'S PLENTY OF ROOM AT THE BOTTOM

In 1959, Feynman delivered a now iconic lecture entitled 'There's Plenty of Room at the Bottom'. In this lecture, he focused on miniaturization and all the obstacles and opportunities it brings. He vividly made his point by asking if it would be possible to write the (then merely) twenty-four volumes of the *Encyclopaedia Britannica* on the head of a pin. The answer is: yes. You can calculate the size of a pin head. You can calculate how many atoms are needed per letter – say, a thousand. But at some point, there is a limit: the quantum limit. Just as there are limits to growth, there are limits to shrinkage. One atom per letter? Far too few. It is precisely this quantum limit that chip manufacturers encounter in their endeavour to shrink transistors to ever-smaller dimensions. And that limit lies in the realm of the nanometre. So? As we've said before, what you can't avoid, you should embrace.

Experimental physicists Monika Aidelsburger (b. 1987) and Immanuel Bloch (b. 1972) share the same vision: building a universal quantum simulator – a quantum system fully under their command. Their dream is to be able to tune every parameter of the system, mimicking any other quantum many-particle system. Their quantum system of choice? The Bose–Einstein condensate. Their first big goal? To replicate the ground state of the two-dimensional Fermi–Hubbard model, the simplest Hamiltonian that should, in theory, explain the essential properties of high-temperature superconductors, where classical simulation methods fail to do so. To simulate this model, they use lasers to create a square optical lattice, a kind of mountain-and-valley structure resembling a big egg carton, with atoms trapped in the valleys. Just like in real high-temperature superconductors. Only via quantum tunnelling do the atoms occasionally dare to hop to another valley, those eggheads.

In an optical lattice, atoms hop from valley to valley.

The main difference between the real-life high-temperature superconductors and their simulation is that, in reality, it is the electrons that jump, while in the simulation, it is the atoms. But this distinction is irrelevant. Despite working on a different scale, the underlying formulas and symmetries remain identical; both types of particles exhibit exactly the same behaviour. This is a textbook demonstration of universality, and the power of the renormalization group.

And this is where things get interesting: in the simulation, certain parameters can be 'tuned'. This is not possible in high-temperature superconductors, where the parameters are fixed by the intrinsic properties of the material. You can't modify the distance between the atoms. A simulation, however, opens up far more possibilities. For example, it allows you to use laser light to change the depth or the distance between molecules and then see what happens. By tweaking these parameters, you can induce phase transitions; you can figure out very clearly how far a system can be pushed before it shifts into a different phase. And it's precisely from these phase transitions that you can deduce which symmetry is breaking – because, ultimately, that's what it always comes down to.

Why is this so thrilling and significant? Because cracking the mechanism behind high-temperature superconductors would clear the way for superconductors that work at room temperature. Such technological progress would be nothing short of disruptive. Everyone can easily imagine a handful of devices that would use next to no electricity.

9.3 Quantum information

But things don't stop there. Beyond sensors and simulators, quantum systems offer a way to tackle fundamental problems in information technology and computing. Among these, two 'wonderful problems' stand out: quantum computing and quantum communication. Let's start with quantum communication. And here we have both good and bad news.

First the bad news. The entanglement patterns of quantum many-particle systems open up an entirely new world of applications for processing and transmitting information, theoretically in an exponentially much more efficient way than is possible with classical systems. So far so good. But there's a catch that tempers this optimism:

a system of n qubits can only store n classical bits of information. This limitation is also known as the 'Holevo bound', named after Alexander Holevo (b. 1943). If you were hoping to use a quantum system to store classical information more efficiently (why not, given the Hilbert space is so absurdly huge?) you are in for a disappointment. It's not going to happen. Quantum systems simply do not have the capacity to store classical information any more efficiently than the average computer.

Why not? Let's say we make a very large superposition, where each element or branch of the superposition represents a page of the massive *Encyclopaedia Britannica*. The problem arises the moment you try to retrieve a single nugget of encyclopaedic wisdom. There's no way to read all the superpositions at the same time. If you read one, the rest of the information instantly vanishes. Because measurement destroys the superposition. Conclusion: it's not possible for a quantum system to store exponentially large amounts of classical information. As soon as you observe the quantum system (or, in this analogy, read one page of the encyclopaedia), the rest of the information is irretrievably lost. This fundamental limitation also means that quantum computers are not likely to boost the development of better algorithms for artificial intelligence and machine learning (such as ChatGPT). Because, in the end, all of these applications run mainly on classical information and they have one common, critical shortcoming: they have very little structure. Whereas quantum has a pronounced 'autistic' side; it absolutely does need structure, something that can be tackled with interference. If you can reduce a problem to a wave-like structure, quantum computing excels. If not, it's simply not quantum's domain.

Now for the good news. The quantum world comes with a range of limitations that don't occur in the classical world. An example: you can never determine both position and momentum of a particle at the same time. You can't perform a measurement with impunity without radically affecting the system. But as counterintuitive as this may sound, these physical limitations open up new possibilities. They allow us to implement new encryption mechanisms, known as quantum cryptography. Here, the security of information transfers is based on the laws of physics, rather than on some 'computational assumption', as in classical cryptographic protocols.

The goal of cryptography is simple: to exchange secret messages between two parties. Let's call in Alice and Bob again. Imagine Alice wants to send Bob a secret message – her highly coveted recipe for Camembert baked in the box. No one else should be able to read it. What's the best way to ensure that? By encoding her message as a string of bits: 1001. Now suppose Alice and Bob manage to create a random secret key that no one else knows, say 0101. Creating such a key is the central goal of cryptography. Using this key, they can encode the preceding message by adding up all the bits bit by bit: 1001 + 0101 = 1100. Alice then sends the message 1100 to Bob. If someone intercepts the message, it's of no use to them – just a string of random bits. But Bob is perfectly able to decode the message. How? By adding the secret bits again bit by bit: 1100 + 0101 = 1001. And just like that, Bob has successfully decoded the message: Camembert baked in the box. Yummy!

How does quantum physics help in creating this key? By allowing Alice and Bob to share a maximally entangled Bell state consisting of two qubits. Here's why. If Alice measures her qubit and gets outcome 0, then Bob is guaranteed to get the opposite outcome – if he measured the same 'observable' as Alice. The measurement results are always entirely random, yet perfectly correlated. The security of this protocol relies on a key property of entanglement: two qubits in a maximally entangled state cannot be entangled with anything else. This behaviour, also known as quantum monogamy, ensures that no eavesdropper can gain information about Alice and Bob's measurements, as the qubits are exclusively linked to each other. And this is precisely the goal of cryptography: to generate pairs of perfectly random and secret bits. With these bits, information can be securely encoded, transmitted and decoded, even over a non-secure channel.

In an alternative version of this cryptographic quantum protocol, a similar effect can be achieved by sending qubits (such as polarized photons), without relying on entanglement. The information of these qubits is encrypted in a randomly chosen base, such as the $|0\rangle$, $|1\rangle$ base or the $|+\rangle$, $|-\rangle$ base (as seen in the Stern–Gerlach experiment). According to Heisenberg's uncertainty principle, an eavesdropper cannot intercept any information unless they know the correct base. This makes it impossible for them to copy the qubits in transit, a limitation known as the no-cloning theorem. You can't measure (or steal, or 'borrow') quantum information without disrupting it (i.e. breaking the superposition). If someone tries to do so, errors will inevitably creep into the communication. Detecting these errors is a sure sign: rivals on the field! In short, quantum cryptography doesn't rely on computational complexity for security – it relies on the fundamental laws of physics.

ENCRYPTING WITH PRIMES

The cryptography used on the internet today is based on a system called RSA (from the initials of its trio of inventors: Rivest, Shamir and Adleman). When encrypted, information sent over a network should be secure, meaning that even if a third party intercepts the message, they shouldn't be able to decode it – unless they hold the right key. RSA exploits the difficulty of factoring large numbers in order to send messages securely: the concept of 'security' is based entirely on 'computational

> assumptions' – the belief that factoring very large numbers is extraordinarily difficult. Factoring means finding the prime numbers by which a number can be divided. For example, the prime factors of 143 are 11 and 13. If you factor 210, you obtain 2 . 3 . 5 . 7. What makes RSA secure is that, as the length of the key increases, the complexity of factoring/cracking it increases exponentially. At least, that's true for a classical computer.

Quantum cryptography is steadily evolving into a mature technology. In fact, there are already commercial systems on the market that leverage this approach. The biggest challenge to widespread adoption lies in the difficulty of transmitting qubits over long distances. Achieving this requires a quantum computer capable of encoding the qubits using a 'quantum error correction code'. We talk about this in detail later.

9.4 Quantum complexity

During the Second World War, while the Manhattan Project was quietly taking shape under the guidance of male scientists, an all-female team was working on another invention of some size aimed at accelerating the incredibly challenging computations required to simulate neutron trajectories. And that 'of some size' applies quite literally: one of the earliest computers weighed a staggering thirty tonnes, housed around 19,000 vacuum tubes, thousands of switches and hundreds of thousands of electrical resistors. It guzzled a hefty 200 kilowatts of electricity and, perhaps most astonishing of all, didn't even have a manual. Compared to modern computers, its memory was roughly that of a goldfish. This Electronic Numerical Integrator and Computer (ENIAC) was maintained by a team of women. Their job? Ensuring that every vacuum tube was interconnected and functioning correctly. In other words, they were the world's first programmers. The computer is a splendid example of emergent behaviour. Thousands of people working on thousands of components that collectively create a functioning whole. Only von Neumann truly understood the entire machine, but every woman was the master of her specific part of the assembly line.

The most transformative shift in the belly of the computer came with the switch from analogue to digital. Either there was power (1) or there was no power (0). From that point on, computers only got faster, smaller and more sophisticated. Over time, vacuum tubes gave way to transistors, which were eventually replaced by microchips. Those microchips, in turn, have dutifully followed the relentless exponential march of Moore's Law, a bold prediction that the number of transistors on a microchip doubles (doubles!) every two (two!) years.

The flip side is this: the faster and smaller we go, the more quantum effects come into play. Classical computers weren't designed for chips and transistors on this minuscule scale. This is why today's computers aren't significantly faster than those from, say, a decade ago. We've hit a boundary, a point where the classical certainty of 0 and 1 starts to blur. But this isn't necessarily a problem; it simply calls for a shift in perspective. If you address this issue through the lens of qubits, a wealth of new possibilities emerges. Naturally, such a fundamental shift in technology means that every aspect of computing must be reimagined. Everything, from hardware to software, programming languages and even the methods programmers use to tackle complex problems, must transform.

But first and foremost, what exactly is a quantum computer? From a theoretical perspective, it is the quantum counterpart of the von Neumann architecture of the classical computer. A quantum computer consists of n different qubits that together form a quantum state. This is the computer's memory. The intention is to entangle those qubits with each other. Whereas classical computers run on algorithms, quantum computers use quantum algorithms. A quantum algorithm is a sequence of quantum operations consisting of 1-qubit operations, and (the much more interesting) 2-qubit operations. The 1-qubit operations are transformations performed on each qubit individually and can be represented as rotations that change the axis of the spins. The 2-qubit operations are obtained by getting two qubits to interact with each other. These interactions can be represented as rotations in a high-dimensional sphere. Once the quantum algorithm has completed a series of 1- and 2-qubit operations, ensuring that each qubit has interacted several times with other qubits, some of the qubits are measured. The outcome of each of these measurements contain bits (classical information: zeros and ones), allowing us to read the result of the computation.

A quantum computer becomes truly advantageous when the total number of 1- and 2-qubit operations required to solve a problem is exponentially smaller than the number of operations needed for a classical computer to perform the same task. The gain in efficiency of a quantum computer over a classical computer lies purely in the fact that, in a quantum computer, the quantum states are at some point entangled. This allows a quantum computer to tackle multiple problems simultaneously, exploring different 'paths' in parallel. The kinds of problems where quantum computers shine are those where these parallel paths can constructively interfere with each other. This is only feasible if the problem has significant structure or symmetry, traits that are often found in quantum many-particle problems.

A WORLD MARKET FOR FIVE COMPUTERS

Strikingly, the first applications for ENIAC were cracking codes and solving physics problems, while the first applications for quantum computers are cryptography and simulating quantum systems. Suspiciously similar. In the early days of classical computing, few could have imagined that, less than a century later, most of humanity would be glued to screens. Back then, no one foresaw the countless ways computers would reshape our lives – like attending live concerts only to view them at home. Even Thomas Watson, the legendary head of IBM, famously misjudged the potential, predicting in 1943 that there might be a global market for, at most, five computers. So if today we can't quite grasp the full potential of quantum computers, it's likely for the same reason: we underestimate human creativity. And perhaps because, at this stage, we can't know that yet. No problem there. As history shows, experiments with unpredictable outcomes tend to be the most interesting. What we do know for sure is that quantum computers will solve problems that classical ones never could.

Feynman had a dream. A quantum dream. He envisioned a quantum computer that could serve as a universal laboratory, capable of performing every experiment imaginable, from anywhere in the world, and doing it much more efficiently than classical computers ever could. Good news for Feynman: on paper, this dream is entirely within reach. All interactions in nature, described by Hamiltonians, are local. This allows the time-dependent Schrödinger equation to be solved efficiently using 1- and 2-qubit operations on a quantum computer. And as the renormalization group has shown us, simulating a quantum system is an intrinsically robust problem. Small, irrelevant errors simply don't matter.

The most compelling applications for such a universal quantum simulator lie in chemical processes, which are a thoroughly quantum mechanical phenomenon and notoriously hard to simulate using classical computers. At present, chemical processes can only be explored in a laboratory setting, which is laborious, prohibitively expensive and occasionally dangerous. A quantum computer could revolutionize this. It could effortlessly simulate chemical processes for pharmaceutical innovation. Or enable research into developing more efficient catalysts and solar panels.

FROM HABER TO HELL

Researchers have explored the feasibility of using a quantum computer to develop more efficient and eco-friendly alternatives to the Haber–Bosch process, a method responsible for a staggering 1.4 per cent of global CO_2 emissions. The Haber–Bosch process is essential for the production of ammonia, the basic component of synthetic fertilizers. It played a pivotal role in averting mass famine in the early twentieth century. Astonishingly, half of the nitrogen in our bodies originates from this process. Yet, there's a darker side to this narrative. The same (Jewish) Fritz Haber (1868–1934) who co-developed this life-saving method also invented mustard gas (or yperite), responsible for some 100,000 deaths during the First World War, and Zyklon-B, with its infamous and devastating legacy.

Quantum computers can solve certain complex questions in an instant. What seems hopelessly intricate for a classical computer might be surprisingly manageable for its unpredictable quantum sister. But what exactly do we mean by complexity? What is it about some problems that makes them so challenging?

In mathematics, computational problems are ranked according to their level of difficulty. The question is how this difficulty (measured as the number of operations required) scales with the length of the input. A classic example is factoring, as mentioned earlier. The length of the input is the number of digits that make up the number to be factored. For a classical computer, factoring is a nightmare because, even with the best classical algorithms, the number of operations needed to factor a number increases exponentially with the number of digits in that number. A quantum computer can do this much more efficiently. Instead of an exponential increase (2^n), it scales polynomially (as a third power, or n^3). So complexity is relative. Whether a problem seems insurmountable depends on whether you're thinking classically or quantumly. And this distinction has profound implications for both science and society. A quantum computer equipped with just a few thousand qubits could theoretically bring the entire internet to its knees – but not before reading all your emails and draining your bank account. This exponential speed-up makes all the difference when dealing with these colossal numbers (with 1,024 digits!) used in RSA encryption, which secures most online communication today.

Another example of a problem with exponential complexity is the 'travelling salesman problem' or its sibling, the 'knapsack problem' (e.g. where you have to pack a supermarket trolley as efficiently as possible). When a delivery driver has to visit a long list of addresses, he naturally wants to know the shortest and quickest route to cover them all. Here, the input length corresponds to the number of parcels or stops. These problems fall under the complexity class 'NP-complete', some of the toughest challenges known. Nobody expects quantum computers to solve such problems efficiently, as they lack the structure quantum systems thrive on. The computer's rigid and structure-dependent nature means it's of little use here. For these cases, brute force is the only option: trying every possibility and seeing what works best. And for the record, some problems are not just hard, they simply cannot be computed, sorry . . .

Back to factoring. We argued that this problem can be solved efficiently using a quantum computer. But how exactly does it work? In 1994, Peter Shor (b. 1959) cracked the code in what could be described as the delivery room of the second quantum revolution. His discovery was rooted in two rather remarkable insights: one purely classical, the other purely quantum. Let's start with the classical part. Shor discovered a hidden structure, a symmetry, within the factoring problem. Suppose you have an enormous number that needs to be factored. Shor's approach involves constructing a periodic function based on that number – a function that repeats itself in a predictable cycle. He demonstrated that the period of this function contains all the crucial information needed to determine the prime factors of the number. Find this period, and you can factor the number. When we talk about a 'period', it naturally brings to mind waves. And for a mathematician, the next step is almost instinctive: apply Fourier analysis (throwback to Chapter 1) to uncover that all-important period.

Peter Shor

Shor's second quantum insight was the fact that this Fourier analysis can be performed exponentially faster on a quantum computer than on a classical computer. With this, he conclusively demonstrated that a quantum computer is capable of factoring large numbers exponentially faster than any classical computer ever would. And thus, it can crack RSA encryption.

But Peter Shor isn't just about algorithms and mathematical wizardry. He also writes less complicated things. Like poems. Take, for instance, his 'Quantum Computer Poetry for the Skeptics', his witty 'poem rebutting a myth about what makes New York City bagels so good', or, in keeping with today's theme:

> *Quantum computers may at first sight seem*
> *To be impossible. How dare we dream*
> *Of solving problems that else would take more time*
> *Than has passed since the cosmos's Big Bang!*

Peter Shor brings us to the second of the four milestones in the history of quantum computing. Here's a quick review.

It all started with the sweary Richard Feynman, who had been the first person to think seriously – albeit in lofty, abstract terms – about the quantum computer capable of running every conceivable quantum experiment. Next came Peter Shor, who proved that quantum computers can factor numbers exponentially faster than their classical counterparts. All it takes is a few qubits and a carefully mixed cocktail of 1- and 2-qubit operations.

Then progress hit a stumbling block. Enter the naysayers, who loudly proclaimed that quantum computers were pointless, nonsensical and destined to fail. Why the fuss? Because many, not entirely incomprehensibly, were daunted by two major and seemingly insurmountable obstacles. First of all: quantum computers operate in the exponentially large Hilbert space, consisting of an equally exponential number of energy levels. Surely, sceptics argued, such a machine would require an exponentially precise control system, composed of utterly utopic lasers. And then came the other problem: decoherence. This merciless phenomenon reduces entangled quantum states to classical states due to interaction with the environment.

9.5 Quantum computers

A quantum computer

In 1995, Ignacio Cirac (b. 1965) and Peter Zoller (b. 1952) founded a club of clear-headed thinkers who did take the quantum computer seriously. While Peter Shor, with his similar aspirations, was already an unofficial member of the movement, Cirac and Zoller were determined to make their mark. With remarkable ingenuity, they demonstrated that implementing a quantum algorithm in a physical system doesn't require exponential control at all. Really? How so?

Here is how. Start with a special kind of chip. The chip is surrounded by ions (atoms with a deficit of electrons), with the two lowest energy levels of each ion acting as a qubit. The 0 of a qubit corresponds to the lowest energy level, while the 1 corresponds to the first excited one. To create a superposition of these energy levels, you simply shine a laser onto the ion, tuned to the energy difference between the two levels ($E = h \cdot v$). This allows control over each individual qubit. However, a quantum computer needs lots of qubits, and they must all be entangled. Because without entanglement, the quantum computer is little more than an elaborate paperweight. So, how can you make qubits interact with each other? Suppose you have a row of ten ions, and you want to make ions two and seven interact. You use a laser to lift each of them individually. Thanks to the electromagnetic forces, they become isolated in a different channel. Then, by making them vibrate, they interact through the phonons and become entangled. Once that's done, you return them to their original spots in the row. Simple as that. On to the next pair. With this ingenious method, Zoller and Cirac demonstrated that the problem wasn't really a problem after all. The precision of control doesn't need to grow exponentially with the size of the problem. It's simply a matter of isolating the right qubits.

This 1995 discovery clearly set the wheels in motion. It was a veritable wake-up call for all quantum scientists, leading to a cascade of new designs and systems aimed at making quantum computers a reality. A recurring question throughout their research is: what type of qubits should you use, and how will you link them?

To date, a wide range of solutions have been proposed, but it's still uncertain which will stand the test of time. Among the frontrunners that are widely adopted and copied are superconducting qubits and silicon spins. Their popularity stems from the fact that their production can fall back on technologies and expertise developed for classical microchips, which makes them highly practical. They also come with the advantage of scalability, are relatively resistant to noise and are highly compatible with modern semiconductor technology. It's no surprise, then, that tech giants like Google, Amazon, Microsoft, Intel and IBM are betting heavily on them. Recently, new competitors have entered the scene, with the neutral atom quantum computer design (based on Rydberg atoms) standing out for its clear advantages in robustness and scalability.

For the time being, quantum computers operating on a thousand qubits – an absolute minimum for tackling really interesting problems – are still a distant dream. The technological hurdles involved in making so many qubits interact are enormous. For perspective, processors in classical supercomputers operate on just 64 or, at most, 128 bits simultaneously, with the rest being 'shuttled' back and forth to memory. But building a similar 'shuttle bus' for qubits is far from straightforward.

The world's largest quantum computers today operate with roughly a hundred physical qubits. However, this is still light years away from achieving a hundred logical qubits. The number of logical qubits dictates the complexity of the problems a quantum computer can solve. A single logical qubit comprises a substantial number of physical qubits because these are highly susceptible to 'noise' (or decoherence). Hence the need to introduce a lot of redundancy. We'll look at this in detail in a moment.

In recent years, the popular science media has flooded us with reports of some or other breakthrough in quantum computing. Isn't that at odds with the previous statements? Indeed, it is. Google's claim in 2019 that their quantum computer of 53 (physical!) qubits performed an experiment that would take a classical computer millions of years was debunked shortly afterwards. The experiment was simulated using a classical computer using tensor networks (a hobby-horse of yours truly – Frank), in mere hours. Similarly, IBM's assertions that their quantum computers are being used to optimize traffic flows are technically true but misleading. Classical computers can do that equally well.

That being said, a future quantum computer with a thousand logical qubits would be capable of tackling some enormously compelling problems. Except we're not quite there yet. The most pressing challenge remains to create an experimental platform that is truly scalable.

And what if, muses the devil's advocate, the quantum computer never materializes, despite the lofty expectations and the astronomical investments? What would that mean for physics? First of all, quantum computing is still in the fundamental research phase. The most significant achievement so far has been the generation of brand-new insights into the many-particle problem, *the* central problem in physics. But it also allows us to ask entirely new fundamental questions. Questions about the role of information in quantum physics, for example. Or about the structure of ground states in systems with topological order. Or about the mysteries of quantum gravity. Or about the role of entanglement in the vacuum of a many-particle system. That's already an impressive list.

9.6 Quantum in error

What is the biggest challenge in building a quantum computer? Ensuring that it stores and processes all those delicate new forms of information (qubits) just as reliably and smoothly as an ordinary computer handles its trusty ones and zeros. Classical computers, after all, are remarkably robust. Their calculations rarely go awry. A 0 can't simply turn into a 1 halfway through a computation. Unless, of course, a rogue cosmic ray decides to interfere. That's exactly what happened in 2003, when the Belgian municipality of Schaerbeek briefly became the epicentre of global news. During the national elections, Maria Vindevoghel of the Workers' Party inexplicably received an extra 4,096 votes – more than the total number of eligible voters. Oops. Coincidence? Hardly. 4,096 happens to be 2^{12}. A wandering cosmic ray had, quite innocently, flipped the thirteenth bit in the electoral computer's memory! A little further afield, say aboard a space shuttle high above the atmosphere, this sort of thing happens all the time, and on a much larger scale. Out there, unprotected classical computers would go haywire as those pesky cosmic rays gleefully flip bits left and right. So, whether in space or in election machines, one thing is clear: error correction isn't just useful – it's essential. But what exactly does that error correction entail?

CLASSICAL ERROR CORRECTION

Error correction owes its origins to the brilliant mind of the aforementioned John von Neumann. He demonstrated that, even with errors persistently sneaking into calculations, it's still possible to perform arbitrarily long computations correctly. Logical bits carry out calculations (as in algorithms). Physical bits, meanwhile, act as the digital filing cabinets, storing this information across multiple locations. But physical bits are fallible. They can be flipped or corrupted, causing errors to creep into the stored data. Fortunately, von Neumann provided a clever workaround: redundancy. By using more physical bits than logical bits and duplicating the data, the system becomes error-resilient. Essentially, the information is spread across multiple places simultaneously, allowing errors to be identified and corrected when they occur.

For instance, let's represent a logical bit 0 as 000 and a logical bit 1 as 111. Imagine an error flips the second bit in each sequence, turning them into 010 and 101. By majority rule, the system identifies that zeros dominate in the first sequence and ones in the second, deducing the original states were 000 and 111, respectively. In essence, more bits are stored than strictly necessary, because zeros and ones are relatively easy to copy. The so-called 'redundant' bits aren't redundant at all; they contain the vital clues needed to detect and fix errors.

The question now is how to incorporate this system of error correction into a quantum framework. Transitioning from classical to quantum computing requires a complete reimagining of error correction, because there are several obstacles.

First of all, isolating a quantum system (qubits) from its surroundings is much harder than isolating a classical system (bits). Ideally, qubits should only interact with one another, but in reality, they constantly interact with their environment. This is a serious issue, because such interactions cause qubits to become entangled with uncontrollable external factors, leading to decoherence (noise). Decoherence breaks the superposition, resulting in a loss of information – an existential problem for quantum computers, as they rely entirely on these superpositions.

Secondly, qubits are inherently more complex than bits. A qubit is not simply 0 or 1; instead, it represents a continuum of possibilities in superposition. This complexity means that quantum systems are susceptible to far more types of errors than the occasional flipping of a 0 to a 1 (or vice versa) in a classical computer. Thirdly, classical error correction relies on redundancy, where information is copied. However, quantum systems are governed by the no-cloning theorem, which states that it is impossible to copy a quantum state.

Finally, detecting an error requires measurement. For classical computers, this poses no problem. But in quantum systems, measurement irreversibly collapses the quantum state, breaking the superposition and destroying the information.

Oh no, laments Jacques the fatalist. Oh yes, exclaims Peter Shor, triumphantly. And he was proved right. Shor tackled these four seemingly insurmountable obstacles and found solutions. He drew support from the idea that the Hilbert space grows exponentially as we add more (physical) qubits. This allowed him to build in redundancy by encoding each logical qubit using many physical qubits arranged in highly specific entangled states. As in Preskill's quantum book (see Chapter 5), each individual physical qubit in itself contains no information about the logical qubit. Instead, the information is entirely encapsulated in the correlations *between* the qubits. The vastness of the Hilbert space provides enough room to encode the same logical qubit in multiple overlapping ways, ensuring that even if some physical qubits are lost or disturbed by their environment, the logical qubit remains intact. All it takes is examining the correlations between the surviving qubits to reconstruct the original state. The same principle applies to error detection. Measuring individual qubits does not reveal any information about the logical qubit itself. Instead, it only indicates whether an error has occurred. Once detected, the error can be corrected without disrupting the overall superposition. In this way, the delicate quantum state remains intact.

And so, in 1995, Shor introduced the first error-correcting code for a quantum system. Initially, these codes addressed only single-qubit errors, but their scope quickly expanded. A growing community of scientists joined Shor, devising increasingly sophisticated schemes and demonstrating that quantum error correction can also be applied to arbitrarily large systems.

TRY SPINNING THAT TO THE CAT

One of the most striking and fundamental discoveries to come out of the study of quantum correction is the existence of a 'quantum threshold'. If every operation in a quantum computer can be performed with a degree of precision exceeding this threshold, it becomes possible to maintain an (encoded) superposition of any number of qubits for an indefinite period. From a philosophical perspective, this is too crazy for words. It suggests that, in principle, nothing prevents us from keeping a cat in a superposition of being both alive and dead. For eternity. At least on paper.

So how is all this even possible? Why does nature allow it? The underlying principle was discovered by Alexei Kitaev (b. 1963): the principle of quantum error correction is equivalent to the existence of anyons (those 'quasiparticles' from the quantum Hall effect) in highly entangled quantum systems. Shor's quantum error correction codes, in essence, emulate anyons. To underpin this dazzling stroke of mathematical brilliance, Kitaev drew on topological field theory, the very same theory that forms the foundation of string theory. This new line of thinking led to two potential pathways for implementing quantum error correction on a large scale. The most elegant approach involves building a quantum computer that naturally hosts anyons. In the year 2000, Microsoft embarked on a major initiative to create such anyons, initially in quantum Hall samples and later in superconductors, where the anyons are referred to as Majorana particles. Yet, these endeavours have borne scant fruit so far. Why? Because getting a handle on anyons is fiendishly difficult – *troppo complicato*. Anyons are proving to be as elusive as Ettore Majorana himself. Born in 1906, Majorana surely has a date of death as well, except we don't know it. This Italian physicist who predicted the Majorana particle vanished at the age of thirty-two. He boarded a ship in Palermo and was never seen again . . .

The 'engineer's approach' to quantum error correction is less elegant – but it works. This method involves simulating a system of (artificial) anyons using a system of ordinary qubits. It is much simpler and is already being employed by tech giants like Google and Amazon.

But if solutions exist, why is it so hard to build a quantum computer? The main obstacle is the massive overhead (the extra effort it requires to obtain a certain result) these solutions entail. If it takes about 1,000 physical qubits to encode a single logical qubit, then a system with just 1,000 logical qubits would require a staggering 1 million physical qubits. This raises a crucial question for the Society of Ingenious Engineers and Ambitious Amateurs: how can we develop better (i.e. more compact) quantum correction codes?

Regardless of the outcome of the quantum computer quest, and whether it ultimately comes to fruition or not, this field of research has already achieved something remarkable: it has united many disparate areas of physics under a common banner. Entanglement, emergence, string theory, anyons; concepts that once seemed abstract and disconnected are now becoming tangible, with practical applications emerging in an ever-growing number of fields.

9.7 Quantum 2.0: entangled particles

Between the first quantum revolution (1925) and the start of the second (1995) lies a vast expanse of seventy years whose passing has wreaked total havoc in physics. It's hardly surprising, then, that today we view ideas and innovations from a century ago through an entirely different lens. Concepts that once seemed outlandish now verge on the ordinary. Just think how quickly the internet, the ubiquity of screens, solar panels or even self-cleaning coffee machines have become part of everyday life. And even since the start of the second quantum revolution, we've already witnessed extraordinary growth spurts.

With each (r)evolution, scientists are compelled to rethink their entire frameworks. Every major leap forward demands fresh mathematics, novel logic, new approaches and, not least, a complete recalibration of intuition. To truly unlock the potential of quantum simulators and quantum computers, researchers must develop a language that abstracts away from the messy details of physical quantum systems. After all, computer scientists creating the software to translate your emails don't need to understand the intricate band structure of silicon. Similarly, to harness the full promise of quantum systems, we need a universal language, one that encompasses all quantum systems and describes them in a unified way. That language is the language of qubits and entanglements. New phenomena demand new words for new terminology. But let's take a step back. Over a century ago, the first strides were made towards understanding nature in terms of *information*.

ENTROPY AND INFORMATION

In the late nineteenth century, three scientists pondered very deeply on the mysteries of statistical physics. They uncovered an inextricable connection – entanglement, if you will – between many-particle physics and information. Their names:

Ludwig Boltzmann, James Clerk Maxwell and Josiah Willard Gibbs.

Let's concentrate here on Maxwell, as his work takes us on a necessary detour to the most important law in (statistical) physics: the second main law of thermodynamics. Thermodynamics is the study of heat – or, more precisely, the way energy can be converted into something else. In the universe, everything always tends to return to a state of equilibrium. One consequence of the second law of thermodynamics is that heat always flows from hotter regions to colder ones, never the other way round. To put it in technical terms: in a closed system, entropy increases. Hot and cold will blend until every particle reaches the same temperature. Entropy can be likened to the disorder that arises when you pour milk into a cup of coffee, or let a drop of ink disperse into a glass of water. The molecules are constantly in motion, jostling and rearranging themselves. Since there are vastly more ways for atoms to mix than to remain separate, reversing the process – unmixing the milk from the coffee or the ink from the water – does not happen. This is entropy. And entropy is irreversible.

According to Einstein, the second law of thermodynamics was absolutely and utterly unimpeachable – unlike Newton's laws, which over time were found to have a few speed-related inaccuracies. But no truth is sacred enough to escape scrutiny and so, in 1867, Maxwell had seen fit to rattle that holy grail. He devised a thought experiment featuring a little demon sitting on top of a box. Why a demon? Because nothing embodies mischief and disorder quite like a demon. In physics, a demon is shorthand for everything that can go awry during experiments and anything that defies our intuition. Very Pauliesque.

A　　　　　B

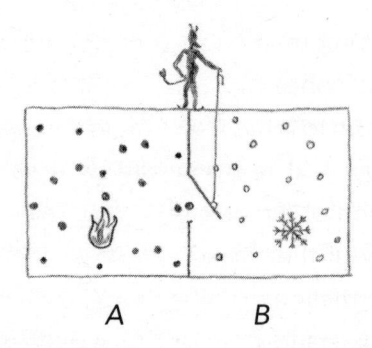

Imagine two sealed chambers, A and B, connected by a partition with a tiny trapdoor. Both chambers contain a gas at the same temperature, with molecules zipping around in constant motion. Maxwell's mischievous demon measures the speed of each particle approaching the trapdoor and deftly flicks the door open, allowing molecules to pass left or right. Fast-moving (hot) molecules are ushered into chamber A, while slow-moving (cold) molecules are herded into chamber B. As time passes, one side grows progressively hotter, the other colder. And herein lies the paradox. Because this process contradicts the famous second law of thermodynamics, which asserts that all particles should eventually reach the same temperature.

Maxwell carried this mystery to the grave, because the solution to this paradox was not found until a century later, thanks to the relay work of three scientists. The first breakthrough came in 1929 with Leo Szilard, who devised a thought experiment demonstrating that information can be converted into energy. In 1960, Rolf Landauer advanced the concept, proving that erasing information inevitably produces entropy. In 1982, the final piece of the puzzle was provided by Charles H. Bennett, a pioneer in quantum information theory and a key figure in the invention of quantum cryptography and quantum teleportation. Bennett solved Maxwell's paradox by simply stating that information and entropy are, fundamentally, one and the same concept.

What it boils down to is this: the demon is not merely opening and closing a door – it is storing information. Every time it measures the speed of a molecule, it records that knowledge. While the entropy of the closed system appears to decrease, the amount of information stored in the demon's brain increases. Therefore, to understand the system's entropy, we must also factor in the demon's mental ledger. And with that, the second law of thermodynamics remains intact! All we need to do is equate entropy with information. And that's kind of surprising, because you would consider information to be an abstract concept, while entropy feels firmly grounded in physical reality.

Physics is, thus, deeply intertwined with the theory of information. John Wheeler (1911–2008), Feynman's PhD supervisor, pushed this idea to its limits with his statement 'it from bit': something only exists if it contains information. Everything is essentially based on, and consists of, the exchange of information. The problem is that no one fully comprehends this idea in its entirety. Even today, the exact connection between quantum physics and information remains an open question. Could there be an information-theorical principle that would unlock the logic of quantum physics? Something that would bring coherence to it all, an underlying principle that explains the laws of quantum physics; a quantum counterpart to the revelation that the theory of relativity stems from the physical law dictating that light travels at a constant speed in any system of reference? It would make Wheeler's prophetic 'it from bit' a reality.

What we do know is how to interpret a quantum state. By now, it's clear that a quantum state is not the description of reality. Rather, it tells us something about the information *we* hold about reality. That's a crucial distinction. This insight forces us to rethink fundamental concepts such as 'quantum state', 'entanglements' and 'measurements'. And, inevitably, this reinterpretation leads to even more questions. But that's the hallmark of healthy research: far more questions than answers. So, let it be known: quantum physics is in excellent health!

A quantum state is essentially a compression, a kind of zip file, of all the information required to predict the behaviour of a quantum system. The double-edged sword of quantum is the fact that these quantum states inhabit a gigantic Hilbert space. The advent of quantum technology cracked open a can of new vocabulary. Out came qubits and entanglements. But can we use these qubits and entanglements to tackle quantum physics more efficiently? Absolutely. To do so, however, we also need a new grammar: rules that define how everything connects and ensure we make sense of it all. Those rules have arrived in terms of quantum circuits and tensor networks. *Quantum circ-who? Tensor net-what?*

'How about,' ventures the professor, '*I* explain these tensor networks and quantum circuits for the aficionados – unfiltered, with no further editing?' A sigh escapes the angelic patience of the writer. 'I'd rather you didn't, but go ahead . . .' (Non-aficionados: feel free to give this a miss.)

TENSOR NETWORKS

The concepts of qubits, entanglements and quantum computing provide a fresh perspective on the central problem of quantum physics: the many-particle problem. What kinds of entanglement are at play here? For both classical and quantum computers, what is the computational complexity of identifying ground states? What is the depth of a quantum circuit with which such ground states can be created using a quantum computer? And does the amount of entanglement evolve continuously or discontinuously when a phase transition occurs?

The first significant insight quantum information theory offered into the many-particles problem is the fact that all physical states, namely the quantum states that can occur in nature, form only a low-dimensional surface ('manifold') in the exponentially large Hilbert space. In other words: the gigantic Hilbert space is an illusion. This obviously leads to the question: what are the properties of all the states in this physical manifold? The most important of these properties relates to ground states of strongly correlated systems, and is called the 'area law of entanglement entropy': the amount of entanglement, or entropy, between one part of the system and its complementary part scales with the size of the boundary between those two parts, rather than with the volume of that one part, as would be the case for an arbitrary state in Hilbert space. This implies that ground states adhere to a kind of holographic principle: all the information is encoded in the boundary.

A second crucial insight was the fact that all ground states of Hamiltonians that are adiabatically interconnected can be transformed into each other using a quantum circuit with a depth that does not scale with system size. This implies that the physical manifold breaks up into equivalence classes of states that can be easily interconverted and, thus, share the same properties. The distinct, disconnected regions of this manifold correspond to the different possible phases. Transitioning from one piece to another requires a phase transition – a process in which a significant amount of entanglement is generated to redraw all the quantum correlations. To create states within 'topological' phases from those in a trivial phase, the depth of the quantum circuit must scale linearly with system size – precisely the property that makes quantum error correction codes possible!

Tensor networks, in two different layouts: PEPS (Projected Entangled Pair State, on the left) and MERA (Multiscale Entanglement Renormalization Ansatz, on the right). At the ends of each edge are two entangled particles (virtual qubits), which are teleported through the vertex (node) throughout the network. The little legs sticking out of the network represent the physical qubits.

The third step involves devising a systematic way to represent states that adhere to the structure of those area laws. This brings us to the concept of 'quantum tensor networks', which are networks of entangled qubit pairs that represent the entanglement properties of the system under study (see illustration). As demonstrated by Steven White, Ignacio Cirac, Guifré Vidal and one of the authors, a crucial feature of these tensor networks is the fact that they offer a completely new way of describing quantum states in terms of (quantum) correlations and (holographic) entanglement degrees of freedom. Here, entanglement takes centre stage, highlighting how the fundamental properties of all states can be far more succinctly represented through the connections between individual particles. The 'entanglement degrees of freedom' provide a holographic image of the state across the entire Hilbert space, while local tensors determine how the entanglement is routed through the system. Qubits and entangled pairs are the new vocabulary of the language of highly correlated systems, and the grammar rules are laid down by the tensor networks.

The main challenge now is to refine the semantics of this new language; a language that describes correlated systems in terms of entanglement unlocks brand-new perspectives on the mystery of many-particle physics. For instance, it enabled us to distinguish topological phases (those lacking local order parameters à la Landau) by examining the different symmetry properties of the underlying tensors. In order for everything to remain the same, everything must change. In this case, the new language reveals emergent symmetries that only become visible in the holographic universe.

Today, tensor networks are a hotbed of global research, yielding groundbreaking variational methods for tackling the many-particle problem. Could it be that we no longer need quantum computers for this task? Tensor networks are also shedding light on new, fundamental aspects of the ubiquitous renormalization group. The long-standing dream of every theorist – to develop a robust and effective renormalization algorithm for highly correlated systems – is steadily moving within reach.

Now that it's clear (to the happy few) what quantum tensor networks are, we can move onto something *really* difficult: how can quantum physics be reconciled with general relativity? It's probably fair to say that every physics student alive secretly dreams of solving this mystery. Unfortunately, no scientist currently has a definitive clue how to proceed. The biggest obstacle is that the general theory of relativity cannot be quantized. And it cannot be quantized because it is not renormalizable – the infinities simply refuse to be tamed. String theory was thought (or, depending on who you ask, is still thought) to hold the key to this solution. After all, string theory elegantly incorporates both quantum theory and gravitational theory, sidestepping those pesky infinities along the way. The only hitch is: it's monstrously difficult. There are a staggering 10^{500} variations of the string theory, and no experiment can tell us which is the right one. So far, string theory has offered precisely zero experimental predictive power, although that's precisely what quantum should be about.

Now you might be wondering – quite rightly – whether we can still call it physics, if you can only use it to prophesize but not to falsify. Regardless, the quest to unify everything under a single theory remains one of the great obsessions of modern science. And it's a perplexing situation: general relativity provides a spectacularly accurate account of the cosmic scale, while quantum takes us deep into the unfathomably small. And yet the two theories are not compatible. Could it be that something is amiss? What are we failing to understand? Have we (yet again) overlooked something?

Juan Maldacena (b. 1968) proposed a potential resolution to the seemingly irreconcilable divide between the two theories in the form of a duality. His idea suggests that quantum theory and general relativity are, in fact, two equivalent descriptions of one and the same underlying theory. Think of it as an orange and its peel: the theory of relativity describes what happens inside the orange, while quantum theory accounts for what goes on in the peel. Both describe the same object, they just offer two completely complementary descriptions of the same, deeper reality. So, perfectly compatible.

And just to be clear, this is pure theory. Hardcore maths. It's far from certain that our universe actually conforms to this model.

As we grope around in the dark, it feels like the perfect springboard to our final leap. Readers who've spent nearly nine chapters eagerly anticipating black holes are finally in for a treat. Everything in the universe is inextricably interconnected yet, as we've hopefully clarified by now, the laws of gravity are light years removed from the laws that govern quantum. Consider a black hole: its position can be precisely pinpointed, and its momentum perfectly known, putting it in blatant defiance of Heisenberg's uncertainty principle.

One of the most famous paradoxes combining quantum and gravity is the black hole paradox devised by Stephen Hawking (1942–2018). The iron-willed Hawking propelled himself to worldwide fame with his prediction that black holes emit a form of radiation, now called Hawking radiation. This discovery, whose characteristics align neatly with Planck's black-body radiation, emerged from calculations that were necessarily part classical and part quantum mechanical. Gravity, after all, cannot be fully commuted in a quantum way. The results of these calculations opened the floodgates to a cascade of other profound questions. What happens to information (not just any information, but the entirety of the now thirty-two-volume *Encyclopaedia Britannica*) when it falls into a black hole? When information crosses the event horizon of a black hole, is it lost for ever? And how does this connect to Hawking radiation? Could the information somehow re-emerge in the radiation emitted by a black hole when it evaporates – a phenomenon that can be taken quite literally and occurs when a black hole radiates more energy than it absorbs? And what does this mean for entropy? The latter is a particularly pertinent question, because the description of a black hole is ultimately about entropy, and entropy is intrinsically tied to information – now identified as a physical entity. Hawking was firmly convinced that something irreversible occurs: what happens in a black hole stays in a black hole. For Hawking, the evaporation of black holes was fundamentally at odds with the generally accepted principle of quantum physics that no information is ever lost. In his view, the foundations of quantum physics needed to be rethought.

The question of what happens to the thirty-two volumes of the *Encyclopaedia Britannica* dumped into a black hole sparked such heated debate that it became the subject of a bet. The bet, between Hawking and John Preskill, revolved around an encyclopaedia of the winner's choice. Preskill, who had long pondered the connection between particle physics and cosmology, was equally captivated by black holes. His perspective, rooted deeply in quantum theory, stood in stark contrast to Hawking's. Nothing, Preskill objected, is irreversible; nothing in nature is ever lost. He was convinced that, sooner or later, and in some way or another, the information sucked into a black hole must re-emerge. How, he couldn't say for sure. Perhaps, if you looked very, very closely, you could read something from the Hawking radiation? While contemplating the bet, Preskill also began to ponder on something else, an idea Richard Feynman, his predecessor at Caltech, had once vividly championed: the quantum computer. Determined to turn this vision into reality, Preskill founded a quantum computing research centre: the Institute for Quantum Information at Caltech. For the record, Preskill's fascination with quantum computing was not driven by practical applications. His sole motivation was to unravel the most fundamental questions in physics. He viewed nature itself as one big quantum computer. And to decode the intricacies of nature, he believed, we must first grasp how quantum computers operate. Perhaps quantum computers will enable us to understand what a black hole really is? Or, more importantly, the role of entanglements in many-particle physics, and what these insights could mean for the development of new materials.

What makes Preskill's institute so exceptional is its ability to bring together a diverse community, all captivated by the allure of quantum information. Black holes, string theory, quantum field theory, emergent phenomena . . . Experts from all these disciplines join forces here and, almost naturally, come to the shared realization that the (new) way of looking at physics based on qubits and entanglements allows them to frame Hawking's yet-unsolved paradox with far greater precision. After all, the clearer a problem is stated, the easier it is to solve.

By 2004 they had taken a first step towards the solution: the information is not lost, it is hidden in the entanglement of the Hawking radiation with the internal degrees of freedom of the black hole. To Preskill's great surprise, Hawking conceded immediately. Preskill found this rather disappointing; their passionate debates, fuelled by their contrasting perspectives, had always borne fruit. Besides, Preskill was convinced the issue wasn't entirely resolved yet. Anyway, he had won, and it was time to claim his prize – *yippie yo yippie yay*! The decision was swift: he opted for a specimen bursting with statistics, figures, comparative studies, dates, achievements and results on an extremely intriguing subject that had completely captured his heart: *Total Baseball – The Ultimate Baseball Encyclopedia*. Would the encyclopaedia of cricket have been just as good . . . ? That was a bit easier to find in the UK (where Hawking was based), you see. You've got to be joking, thought the American, refusing to yield. Though, as Preskill carried his nearly 2,700-page tome home, he couldn't help but concede that it weighed as much as a black hole.

Fast forward twenty years. A second version of the solution popped up, yet it is so full of arithmetical trickery that it's almost as hard to understand the answer as it was to find it in the first place.[1] At its core, it incorporates everything that physics stands for

---

[1] Ahmed Almheiri, Thomas Hartman, Juan Maldacena, Edgar Shaghoulian and Amirhossein Tajdini, 'The entropy of Hawking radiation', *Reviews of Modern Physics* 93(3), 035002 (2021).

today: quantum computing, quantum entanglements, entanglement monogamy, quantum error correction, the holographic universe . . .

All these platonic ideas sparked a short-circuit in the mind of Markus Aspelmeyer (b. 1974), the ever-cheerful and affable physicist from Vienna. Fed up with the endless theories about reconciling the irreconcilable, he locked himself in a lab. Actions, not words! Aspelmeyer's goal: to create a superposition of two nanoparticles, each existing simultaneously in two locations.

Quantum gravity in action. And then? See what happens. How will interference combine with gravity? How do the particles attract each other? Quantum physics offers no definitive answers here because it operates in a static space-time. Or, more accurately: quantum falls short here. The theories and mathematical equations we have simply can't compute the result. The outcome is unpredictable. And so, physicists need solid experimental data – something to point them in the right direction. To this day, quantum physics and gravity remain separate continents, divided by a deep, vast ocean with an infinite horizon in between; mini-nanoparticles versus the grandeur of gravity. Aspelmeyer might be the first to bridge these two worlds, these two theories (because experiments are unencumbered by the insurmountable wall of infinity). He's putting the cat among the pigeons. After all, nothing ventured, nothing gained. On the other hand, maybe there's no problem at all with the non-renormalizability of quantum gravity. Quantum theory works. No one disputes gravitational theory. But perhaps there's something else, something that lies between the two? Maybe it's misguided to believe there is a single theory that unites both worlds? Could it be an illusion to believe there is a theory that describes everything, one that's internally consistent? Or is it simply that our human-centred mathematics isn't up to the task of expressing nature's language? Were Stevin and Galilei overly optimistic? Maybe. But maybe not. Either way, if the theory doesn't hold up, the revolution begins anew!

EPILOGUE

It all ended with a huge party to celebrate a hundred years of quantum physics. The festivities took place in a recently refurbished brewery, set among open fields with views of the sea. The owner was a certain Mr Quanten MacAnnick. The birds chirped (*pi pi pi*), the frogs croaked (*quark quark*) and apart from a few wisps of mist, there was not a cloud in the sky. The party could commence!

By the time the pendulum clock chimed six, the first guests had started trickling in. Although it was all rather chaotic from the off, the atmosphere was almost unbearably cosy. Planck sat at the piano, playing one of his own (very classical) compositions, while Einstein tucked his Lina between the warm folds of his neck. Feynman amused himself on the drums, not caring about anything else. Heisenberg, feeling uncertain, hesitated for a while before tentatively asking Einstein if he would like to play piano four hands with him, but Einstein didn't feel like it. Heisenberg was, in his opinion, on too much of a different wavelength. 'Try Born,' Einstein suggested, 'he's good on the keys too.' Besides, Heisenberg was in a bit of a mood because there was another party going on elsewhere, but he couldn't possibly be in two places at the same time. To make matters worse, he found himself cornered by the ever-chatty Bohr, who wanted to explain something very complicated to him – or wait, was that Born? Ah, they are so indistinguishable . . .

A short while later, Emmy Noether walked in, heavily pregnant (her symmetries had already broken), bursting with new ideas. She rushed over to Stevin, offering him one of her homemade biscuits – the poor old man had suddenly collapsed. Maybe he hadn't been eating his apple a day? But, coincidence or not, Newton had baked apple pie with sugar crystals sprinkled on top, while Oppie had come trotting along with apple tart – only his apples looked a bit toxic. Oppie was a very contagious eater though, it has to be said; all his *nim nim nimming* soon had the entire table *nim nim nimming* along with him. He clearly didn't take part(icle) in elementary manners . . . In any case, sweet indulgence galore. Thomson delighted the company with his infamous plum pudding. Tragically, someone had picked out all the currants (we won't name names, but in all likelihood it was that rascal Rutherford). Meanwhile, Dirac was noting that Schrödinger and Heisenberg, although following completely different recipes, had somehow brought exactly the same appetizers. Then, suddenly, a high-pitched squeal echoed from the sea. It was Newton, leaping in like an overexcited child – he had never seen anything like it before. Einstein quickly captured the moment with his trusty photon camera.

It was time for Maria Mayer's onion soup – which was burned to a crisp, courtesy of that ever-pernicious Pauli. Einstein, naturally, was sulking, complaining that the soup was too hot. Meanwhile, the cat, purring quietly under the table – for a moment it was presumed dead ('That is not just uncertain,' Pauli sneered, 'it is not even wrong!') – suddenly jumped up, causing the troubled Stevin to fall for a second time, just as hard, with his soup bowl still in hand, and Landau was left to pick up the broken symmetry.

Eventually, everyone settled back into their meals, though all eyes turned to Wilson, who was celebrating his happiness with an impromptu folk dance (that wasn't exactly normal either!). They got a bit of a shock, though, when Hamilton entered in three-dimensional slow motion (too much time spent watching *The Matrix*, that one). Behind him, the French prince made his signature grand entrance – regally late, of course, and through two doors at the same time. Had he not brought anything to share? *Non*. Well, he had brought Gamow, that ray of sunshine! Speaking of sunshine, Curie's chicken was a triumph, and she herself was truly radiant. It did not escape anyone's attention that she immediately gravitated toward Rutherford, our alpha male – the chemistry between them was truly contagious.

Oh look, there's Onnes, quick and lively as ever, with Anderson hot on his heels – and clearly having a great time. And that's quite striking, considering he doesn't really get along with everyone individually, but in a group setting he's clearly in his element. 'It's like a fairy tale,' he said. It was indeed no standard party, and that's exactly what made this moment so special, Weinberg confirmed. Seeing everyone united tugged at his heartstrings. This, in turn, rang a Bell with Shor, who sidled up to his friend and deviously carved a few lines of poetry into the table (where had he picked up *that* idea?). Incidentally, Hilbert wasn't invited this time, he took up a bit too much space; it wasn't meant to be that kind of networking event. To avoid unnecessary tension, he'd been politely asked to stay away. Galois was also there, of course, though he was sitting broken-hearted all alone on the ground (states we'd rather leave unspoken, but look, the group theo— I mean, group therapy, would certainly do him good), even though he'd actually been assigned the super job of conducting everyone to their place at the table, but so be it.

Ding ding, would someone like to read the speech? Fermi? No no, he never gets to the core. Schrödinger? Forget it, he's entangled somewhere under the straw with a girl he met in a wave of excitement. Mendeleev then? Hmmm, too much compartmentalization. Anyon . . . ?! Yes, let's ask Einstein! He's an excellent MC^2! At which point Einstein stepped forward, positioned himself resolutely in the LED light and launched into his speech: 'Dearest friends . . .' But just then, he froze – a black hole in his memory. Blushing and searching for help, wishing that the ground would swallow him up, he turned to the wings, where Lise Meitner immediately stepped out of the shadows. She urged Einstein to take it all in relative terms, before taking over the speech herself – but not without giving it an unexpected quantum leap. 'Ladies!' she began, getting straight to the point, and she read out a short text from the first page she had once, in a moment of inspiration, torn out of a book and kept safe in her purse ever since – as if it were her personal mantra:

> *The atoms and their hidden reign,*
> *unseen by the eye,*
> *unknown to the brain.*
> *Our dreams fall short,*
> *they cannot hold,*
> *a vision that no mind can mould.*
> *Quantum, a spark in recent years,*
> *met our thought uncertain and unclear.*
> *Yet now we see, as both align:*
> *a deeper glance*
> *rewrites the line.*
> *It breaks the truths we thought we knew,*
> *revealing worlds beyond our view.*

ACKNOWLEDGEMENTS

First of all, our heartfelt thanks go to Laura Lannoo for envisioning the idea of crafting an accessible book on quantum physics, to Lannoo Publishers for following the unpredictable quantum trajectory of this project, and to David Tong, Peter Tallack and the entire Pan Macmillan team for making the English edition a reality. Amaryllis, your candid reminder that understanding quantum physics is one thing, but explaining it – or distilling it into prose – is quite another, was both humbling and invaluable. Like a *deus ex machina*, Céline cycled past (literally) and, in all her uninhibited craziness, took up the pen. All the beauty in this book comes from Céline, all the awkwardness from Frank.

Thanks to Karel Van Acoleyen for helping to draw up an initial outline of the book – Simon Stevin is watching attentively from the Nieuwe Kerk. Thanks to Koenraad Audenaert and Alexandre Sevrin for their astute observations – just because the book is a popularizing one does not mean it should not be 100 per cent correct. Thanks to Bartel Broeckaert and Gaspar Verstraete for bringing out the best of both the poetry and grammar to the original Dutch edition.

Explicit thanks to the FWO through the EOS project CHEQS for making this project possible; our greatest wish is that this book will inspire young students to become physicists.

Céline, finally, would like to express special thanks to Frank, for being crazy enough to suggest writing this book together. You with your maths. Me with my Dutch. Who would have thought that this would form such a wonderful marriage?

GLOSSARY

ALGORITHM: a set of operations to be performed to solve a mathematical problem.

ALPHA PARTICLE: the nucleus of a helium atom.

AMPLITUDE: the extent to which a wave moves up and down.

ASSOCIATIVE: basic property of group and matrix multiplications. It states that the product of three operations (a, b, c) can be calculated in two equivalent ways, either by first multiplying a by b and then by c, or by multiplying a by the product of b and c.

BAND STRUCTURE: the collection of various energy orbitals (of electrons in a crystal). Within a band, there is a continuum of possible energies, whereas different bands are separated by gaps (forbidden energy levels).

BELL PAIR: a state of two qubits with maximum entanglement.

BETA PARTICLE: a high-energy electron or antielectron.

BOSON: elementary particle with a symmetric wave function. The most well-known boson is the light particle (photon). Derived from its 'discoverer', Satyendra Nath Bose. Counterpart of the fermion.

COMMUTATIVE: a special property of certain group and matrix multiplications. When the multiplication of a by b is equal to the multiplication of b by a, then a and b commute. The most interesting matrices in quantum physics do not commute (they are non-commutative). In those cases (a . b) is not necessarily equal to (b . a). This non-commutativity forms the mathematical basis for the uncertainty principle of Heisenberg.

COMPLEX NUMBERS: extension of the real numbers. Thanks to complex numbers, the square root of negative numbers can also be calculated.

CONTEXTUALITY: the properties of a system depend on the way you look at it (the context).

DE BROGLIE WAVELENGTH: the quintessential manifestation of wave–particle duality. The de Broglie wavelength quantifies the extent to which a particle can be localized (as a wave packet). The lighter a particle and/or the colder it becomes, the larger the wavelength. When the distance between particles exceeds their de Broglie wavelength, classical physics suffices to describe their behaviour. However, when this distance falls below the wavelength, interference effects arise, and quantum becomes essential.

DECOHERENCE: a technical term describing the effects of the environment (and noise) on a quantum system. Decoherence destroys coherence (superpositions). This is the Achilles' heel of the quantum computer.

EIGENFREQUENCY: property of a matrix or a Hamiltonian. The eigenfrequencies correspond to the energy levels of the 'stationary' states or electron orbitals of the system. Also known as natural frequency.

ELECTRON: an elementary particle with a negative charge, responsible for the chemical properties of all matter. Electrons are fermions, and therefore obey Pauli's exclusion principle; their corresponding repulsion gives rise to the hardness and stability of matter.

ELECTRON ORBITAL: the region around the nucleus of an atom where an electron is most likely to be found. These orbitals are determined by solving the (single particle) Schrödinger equation and have specific shapes and energy levels. Some orbitals are spherical (called s orbitals), some are shaped like a dumbbell (called p orbitals), and others have more complex shapes (like d and f orbitals). Orbitals help us understand how electrons are arranged in an atom and how they interact with other atoms.

EMERGENT: property that occurs only in systems consisting of many particles. Many atoms behave in a completely different way to a single atom. Depending on the scale on which you do physics, other organizing principles and laws apply.

ENERGY ORBITAL: see electron orbital.

ENTANGLEMENT: property of the wave function of two or more correlated particles. Responsible for a lot of the weirdness in quantum physics, but also for the computational power of a quantum computer. When two particles are entangled, a measurement on one instantaneously influences the state of the other, regardless of the distance between them.

ENTROPY: a measure of the disorder or randomness in a physical system. The second law of thermodynamics dictates that in a closed system the entropy always increases.

EPR: thought experiment, named after Einstein, Podolsky and Rosen, that examined the properties of entangled particles with the aim of challenging Heisenberg's uncertainty principle.

EXCITED STATE: an electron in an orbital that is not the lowest energy state.

EXCLUSION PRINCIPLE: one of the most important principles in quantum physics, discovered by Wolfgang Pauli. Each electron orbital can hold only two electrons, and they must have opposite spin. It is this principle that ensures the stability and hardness of matter.

EXPONENTIAL: growth where the rate increases in proportion to the current size, e.g. adding one variant doubles the number of variables.

FACTORING: the process of writing a natural number down as a product of prime numbers.

FERMION: particles with an antisymmetric wave function. Examples of fermions include electrons, protons, neutrons and quarks. Fermions repel each other very strongly due to Pauli's exclusion principle. Counterpart of the boson. Named after Enrico Fermi.

FEYNMAN DIAGRAM: a fundamental tool in perturbation theory that simplifies complex calculations, making it possible to describe and understand interactions between particles.

FIELD THEORY: a mathematical theory that assigns a degree of (quantum) freedom to each place in space. It is the only approach capable of unifying quantum mechanics with the theory of relativity.

FREQUENCY: the number of oscillations a wave completes in one second.

FUNCTION: a mathematical expression that describes physical quantities, such as position or energy, across space and time. For example, a wave function represents a particle's quantum state and enables outcomes of experiments to be predicted.

GAMMA PARTICLE: photon with a very high energy/frequency.

GAUGE THEORY: theory in which the symmetry is locally conserved everywhere.

GROUND STATE: the wave function with the lowest possible energy.

GROUP: mathematical formalism to describe symmetries.

HALF-LIFE: the average time it takes for half the atoms in a radioactive material to be 'reincarnated' into another species.

HAMILTONIAN: describes the kinetic and potential energy of particles and the interactions between particles. Named after William Hamilton.

HARTREE–FOCK METHOD: an approximate method for describing the wave functions of many-electron systems by representing them as a combination of single-electron orbital wave functions.

HIDDEN-VARIABLE THEORY: a theoretical framework suggesting that the randomness in quantum mechanics arises from underlying, unobserved variables that determine the outcomes of quantum events. This theory has been debunked by Bell.

HILBERT SPACE: mathematical framework used to describe the complete set of possible wave functions of a quantum system, represented as vectors in an infinite-dimensional space.

INTERFERENCE: a phenomenon arising from the superposition of waves, such as probability amplitudes, where overlapping waves combine to create regions of increased or decreased intensity. This effect underpins many quantum behaviours, such as the double-slit experiment, where particles exhibit wave-like patterns.

ISOTOPE: atoms with the same number of protons or electrons, but with a different number of neutrons.

KINETIC ENERGY: the energy a particle or object possesses due to its motion.

LOCAL REALISM: a philosophical view proposed by Einstein, which states that all properties of particles are fixed, and do not depend on the way we look at them.

LOCALITY: principle that an object is directly influenced only by its immediate surroundings, and any interaction between objects separated by a distance cannot occur faster than the speed of light, in accordance with the theory of relativity.

MATRIX: a mathematical tool used to represent transformations of vectors, consisting of numbers arranged in a grid, similar to a chequerboard. Two matrices can be multiplied, but this operation is generally non-commutative. Groups can be represented by matrices, and the Schrödinger equation can be interpreted as the action of an infinite matrix (the Hamiltonian) on the wave function (a vector).

MATRIX MECHANICS: Heisenberg's version of quantum physics, in terms of matrices.

MOMENTUM: velocity times mass.

n: reference to an undetermined number.

NEUTRON: the neutral counterpart of the positive proton in the atomic nucleus.

NODE: the points where the (wave) function is zero.

NONLOCALITY: phenomenon where particles that are entangled influence each other instantaneously, regardless of the distance separating them.

NUCLEAR FISSION: a process in which the nucleus of a heavy atom splits into two or more smaller nuclei, releasing energy and often additional particles, such as neutrons.

NUCLEAR FORCE: the weak nuclear force is responsible for converting a neutron into a proton and drives radioactive processes in the nucleus. The strong nuclear force binds protons and neutrons together, ensuring the stability of the nucleus.

NUCLEAR FUSION: a process in which lighter atomic nuclei combine to form a heavier nucleus, releasing a significant amount of energy.

NUCLEON: collective name for protons and neutrons.

OBSERVABLE: anything that can be observed and measured. Observables are represented by matrices in quantum mechanics.

PERMUTATION: a specific rearrangement of a set of objects.

PHASE TRANSITION: a transformation in the physical state of a system, such as from solid to liquid, liquid to gas, or between different quantum phases.

PHOTON: an elementary particle that is the fundamental unit, or quantum, of light.

POTENTIAL ENERGY: the energy stored in an object because of its position, and due to its interaction with other objects.

PROTON: one of the building blocks of the nucleus. It is a positively charged fermion. For atoms in the periodic table, the number of protons in the nucleus is always equal to the number of electrons.

QUANTIZATION: the concept in physics where physical properties, such as energy or frequency, are restricted to discrete values rather than being continuous.

QUANTUM: a discrete, indivisible unit of energy or physical quantity. For example, light is not a continuous stream but is emitted and absorbed in quanta, or packets, of energy called photons.

QUANTUM COMPUTER: computer that uses qubits instead of classical bits, leveraging quantum principles such as superposition and entanglement. Its power lies in the ability to perform exponentially many calculations simultaneously, enabling solutions to problems that are infeasible for classical computers.

QUANTUM LEAP: the smallest possible jump one can make.

QUANTUM MANY-PARTICLE SYSTEM: a system consisting of many quantum particles.

QUANTUM MECHANICS: synonym for quantum physics. 'Mechanics' refers to the branch of physics that studies the motion of objects and the forces that cause or affect this motion.

QUANTUM PHYSICS: the branch of physics where the wave-like properties of particles play a fundamental role. Whether you need classical or quantum physics depends on the de Broglie wavelength of the particles in question.

QUANTUM STATISTICS: a framework for describing a collection of indistinguishable quantum particles. Fermions follow Fermi–Dirac statistics, obeying the Pauli exclusion principle, while bosons follow Bose–Einstein statistics, allowing them to occupy the same quantum state.

QUARK: the building blocks of a wide variety of elementary particles, belonging to the family of fermions. They exist in six 'flavours'. The up and down quarks are the most prevalent ones. Quarks always come in pairs or triples.

QUATERNION: a mathematical concept that extends complex numbers, consisting of four components. Multiplying quaternions is non-commutative. Quaternions were the precursor of matrices and feature prominently in quantum mechanics by means of the Pauli matrices.

QUBIT: quantum system in a superposition of two possible states. Building block of the quantum computer.

REDUCTIONISM: a concept that states that the behaviour of the whole can be fully understood by analysing the behaviour of its individual components. Counterpart of emergence.

RENORMALIZATION: a way of describing how a theory changes depending on the scale at which a system is observed. In the context of quantum field theory, renormalization often involves dealing with infinities, which is a major source of its complexity.

SEMICONDUCTOR: a material with electrical conductivity between that of a conductor and an insulator, whose properties can be controlled by temperature, impurities or external fields, making it essential for modern electronics. Basic building block of transistors.

SHELL: collection of the most energetic (partially) filled electron orbitals of an atom. The electrons in this (outer) shell are responsible for the chemical properties.

SPIN: a fundamental, quantized property of elementary particles. In the case of an electron, its spin can take two values (up and down) and is responsible for its magnetic moment. Electrons in the same electron orbital always have opposite spin. The quantization of the spin was first demonstrated by Stern and Gerlach.

SUPERCONDUCTOR: a material that, below a certain critical temperature, conducts electricity with zero resistance and expels magnetic fields, enabling highly efficient energy transfer and powerful magnetic applications.

SUPERPOSITION: a fundamental principle of quantum mechanics, where a particle exists simultaneously in multiple states/positions/orbitals. The superposition principle expresses the wave nature of quantum particles. The double-slit experiment demonstrated that an electron can exist in a superposition, effectively passing through both slits at the same time.

SYMMETRY BREAKING: a basic mechanism for distinguishing phases of matter. During a phase transition of a many-particle system, a symmetry is always broken (or restored).

THOUGHT EXPERIMENT: experiment that is only performed in thought, but not in practice.

TRANSLATIONAL SYMMETRY: the property resulting from the fact that a system looks identical when it is moved (translated) over a certain distance.

TUNNELLING: a quantum phenomenon where a particle passes through a potential energy barrier that it classically would not have enough energy to overcome. This is the fundamental mechanism behind radioactive decay.

UNCERTAINTY PRINCIPLE: certain pairs of properties, such as position and momentum, cannot be precisely measured simultaneously. The more accurately one property is known, the less precisely the other can be determined. It is a consequence of the non-commutativity of the matrices (observables) associated with position and momentum.

VACUUM FLUCTUATIONS: a result of the fact that the ground state of a quantum system is typically a highly entangled superposition of many possible states. Vacuum fluctuations are responsible for all exotic properties of quantum materials.

VECTOR: a mathematical object represented by a row of numbers. For instance, it can describe a direction in space, or represent a wave function in Hilbert space.

WAVE FUNCTION: a mathematical function that describes the quantum state of a particle or system, encoding information about all observables, such as position and momentum. The Schrödinger equation dictates how the wave function evolves over time. The wave function does not represent reality. It does not represent a physical wave, but rather the information *we* have about a system, and enables the calculation of the probabilities of various measurement outcomes.

WAVE PACKET: a superposition of waves with different wavelengths and frequencies, forming a localized 'packet' that describes a particle. See also de Broglie wavelength.

WAVELENGTH: the distance between two consecutive peaks of a wave.

ψ: (*psi*) symbol for the wave function, introduced by Schrödinger.

FURTHER READING

Anderson, Philip W., *More and Different: Notes from a Thoughtful Curmudgeon*, World Scientific, 2011.
Baggott, Jim, *The Quantum Story: A History in 40 Moments*, Oxford University Press, 2011.
Bhattacharya, Ananyo, *The Man from the Future: The Visionary Life of John von Neumann*, Penguin UK, 2021.
Bird, Kai and Sherwin, Martin J., *American Prometheus: The Triumph and Tragedy of J. Robert Oppenheimer*, Atlantic Books, 2021.
Bizony, Piers, *Atom (Icon Science)*, Icon Books, 2017.
Brown, Brandon R., *Planck: Driven by Vision, Broken by War*, Oxford University Press, 2015.
Cathcart, Brian, *The Fly in the Cathedral: How a Small Group of Cambridge Scientists Won the Race to Split the Atom*, Penguin Books, 2004.
Farmelo, Graham, *The Strangest Man, The Hidden Life of Paul Dirac*, Basic Books, 2016.
Farmelo, Graham, *The Universe Speaks in Numbers: How Modern Maths Reveals Nature's Deepest Secrets*, Faber & Faber, 2019.

Feynman, Richard P., *'What Do You Care What Other People Think?': Further Adventures of a Curious Character*, WW Norton & Company, 2001.

Feynman, Richard P., Leighton, Robert B. and Sands, Matthew, *The Feynman Lectures on Physics, Vol. III: Quantum Mechanics*, Basic Books, 2011.

Feynman, Richard P., *The Meaning of It All: Thoughts of a Citizen-Scientist*, Basic Books, 2009.

Galilei, Galileo, *Dialogue Concerning the Two Chief World Systems*, Modern Library Inc, 2001.

Gamow, George, *The Great Physicists from Galileo to Einstein*, Courier Corporation, 1988.

Gamow, George, *Thirty Years That Shook Physics: The Story of Quantum Theory*, Courier Corporation, 1985.

't Hooft, Gerard, *In Search of the Ultimate Building Blocks*, Cambridge University Press, 1996.

Humphrey, Marc, Pancella, Paul V. and Berrah, Nora, *Idiot's Guides: Quantum Physics*, Alpha Books, 2015.

Hürter, Tobias, *Too Big for a Single Mind: How the Greatest Generation of Physicists Uncovered the Quantum World*, The Experiment, 2022.

Ignotofsky, Rachel, *Women in Science: 50 Fearless Pioneers Who Changed the World*, Ten Speed Press, 2016.

Isaacson, Walter, *Einstein: His Life and Universe*, Simon & Schuster, 2007.

Kragh, Helge, *Quantum Generations*, Princeton University Press, 2020.

Laughlin, Robert B., *A Different Universe: Reinventing Physics from the Bottom Down*, Basic Books, 2005.

Moore, Walter, *Schrodinger: Life and Thought*, Cambridge University Press, 1989.

Musil, Robert, *The Man Without Qualities*, Vol. 1., trans. Eithne Wilkins and Ernst Kaiser, Secker & Warburg, 1953.

Orzel, Chad, *How to Teach Quantum Physics to Your Dog*, Simon & Schuster, 2010.

Parisi, Giorgio, *La chiave, la luce e l'ubriaco, Come si muove una ricerca scientifica*, Di Renzo Editore, 2006.

Penrose, Roger, *The Road to Reality*, Random House, 2006.

Raymer, Michael G., *Quantum Physics: What Everyone Needs to Know*, Oxford University Press, 2017.

Rovelli, Carlo, *Helgoland: Making Sense of the Quantum Revolution*, Penguin Books, 2022.

Schwartz, David N., *The Last Man Who Knew Everything: The Life and Times of Enrico Fermi, Father of the Nuclear Age*, Hachette UK, 2017.

Schweber, Silvan S., *QED and the Men Who Made it: Dyson, Feynman, Schwinger and Tomonaga*, Princeton University Press, 2020.

Segrè, Emilio, *From X-Rays to Quarks: Modern Physicists and Their Discoveries*, Courier Corporation, 2012.

Snow, Charles Percy, *The Two Cultures*, Cambridge University Press, 2012.

Ulam, Stanislaw M., *Adventures of a Mathematician*, University of California Press, 1991.

Vanden Berghe, Guido, Viaene, Dieter and Vandamme, Ludo, *Simon Stevin van Brugghe (1548–1620): hij veranderde de wereld*, Sterck & De Vreese, 2020.

Weisskopf, Victor F., *Knowledge and Wonder: The Natural World as Man Knows It*, MIT Press, 1979.

Weyl, Hermann, *Symmetry*, Vol. 47, Princeton University Press, 2015.

Zangwill, Andrew, *A Mind Over Matter: Philip Anderson and the Physics of the Very Many*, Oxford University Press, 2021.

INDEX

Page references in *italics* indicate images.

't Hooft, Gerard 178*n*, 220, 254, 258, 260

Aidelsburger, Monika 299–300
alpha particle 87, 226, 227, 229, 233–5, 237, 243, 263, 264, 342
Alpher, Ralph 263
Anderson, Philip W. 220, 253, 259–60, 268, 271–2, *272*, 275–6, 277, 289, 337
anyon 291–3, *292*, 294, 320–1, 338
Arendt, Hannah 164–5, 165*n*
Aspect, Alain 160
Aspelmeyer, Markus 334
A-team, the 105
atomic clock 216, 296–8

band structure 166, 202–5, *204*, 322, 342
Bardeen, John 166, 204–5, 268, 283, 285
BCS theory 283
Becquerel, Henri 221–2, 223
Bell, John 138, 141, 151, 156–7, 158–61, *159*, 163, 182, 303, 342, 346
Bennett, Charles H. 294, 324–5
Berkeley, George 118

beta particle 234, 237, 243, 246, 257, 263, 342
Bethe, Hans 242, 263
binding energy 80, 118, 191
black emitter 67–8
black hole 84, 216, 240, 271, 331–3, 338
Bloch, Immanuel 299–300
Bohr, Niels 43, 64, 85–6, *85*, *86*, 88–91, *90*, 94, 107, 118, 120*n*, 125, 129, 138, 141, 144, 145–6, 148–51, 152, 155, 156, 164, 167, 211, 227, 238, 242, 243, 244, 336
Boltzmann, Ludwig 72, 95, 173–4, 323
Born, Max 104, 106, 109, 111, 113, *114*, 115, 118, 119, 142, 239–40
Bose, Satyendra 48, 166, 171, 212–13, 342
Bose–Einstein condensates 48, 170, 171, 212, 214, 215, 216–17, 218, 281, 282, 284, 295, 298, 300, 349
boson 45, 48, 166, 170, 171, 177, 179, 183, 210, 211–12, 213–14, 217, 246, 247, 248, 252–4, 257–8, *257*, 259, 261, 271, 281, 284, 291, 342, 345, 349

Brout, Robert 220, 253, 260

calculus 20–5, 45, 51
Carlsberg 70, 91
CERN 247, 254, 259, 285
Chadwick, James 228–9
Char, René 6
Cirac, Ignacio 294, 313, 314, 328
Clauser, John 160
complex numbers 26–9, 107, 108, 109, 113*n*, 124, 169, 343, 349
contextuality 163–4, 343
Cooper, Leon 283–4, 285
Cornell, Eric 215
correspondence principle 120*n*, 121, 125, 210–12
crystals 41–3, 97, 178, 201, 202, 203–5, 222, 283–4, 336, 342
Curie, Marie 144, 220, 222–5, *223*, 230, 238

Davisson, Clinton 97
de Broglie, Louis 61, 64, 92–7, *92*, 92*n*, 99–100, 102, 106, 107, 112, 114, 118, 124, 125, 132, 144, 167, 210–11, 214, 284, 343, 348, 351
Dirac, Paul 91, 104, 106, 120, 121–8, *122*, 144, 171, 176, 248–9, 250, 270, 280, 298, 336, 349
du Châtelet, Emilie 25–6, 36, 245
Dyson, Freeman 241, 251

Einstein, Albert 2, 5, 23, 35, 37, 48, 61, 64, 76–84, *76*, 77*n*, *84*, 85, 87–8, 94, 107–8, 110, 112, 121, 124, 125, 129, 131, 133, 138, 139, 141–52, 155, 156–7, 160, 161, 163, 164, 166, 167, 170, 171, 174, 210–18, 221, 224, 234, 238, 241, 242–3, 247, 249, 250, 270, 281, 282, 284, 295, 297, 298, 300, 323, 335–7, 338, 344, 346, 349
emergence 44, 82, 216, 269–78, 281, 282, 284, 286, 289, 290, 291, 292, 298, 306, 321, 329, 333, 344, 349

Englert, François 220, 253, 259–60
entanglement 8, 110, 115, 140, 141–2, 151, 155, 156, 157, 160, 161, 162, 167–8, 178, 179, 182, 270, 292, *292*, 293, 294, 295, 301, 303, 304, 314, 316, 321, 322, 326–9, 332–4, 342, 344, 348
entropy 182, 322–5, 327, 331–2, 333*n*, 344
EPR 138, 146–51, 152, 155, 344
exclusion principle 2, 48, 54–5, 126, 171, 177–8, 211–12, 265, 343–5, 349

Fermi, Enrico 122, 166, 171, 220, 224, 234–7, *235*, 241–2, 245, 246, 255, 300, 338, 345, 349
Fermi–Dirac statistics 122, 171, 349
fermion 48, 166, 170, 171, 177, 178, 179, 180, 181, 210–12, 246, 252, 254, 257, 261, 291, 292, 342, 343–4, 345, 348, 349
Feynman, Richard 100, 123, 128, 139, 166, 178–80, 178*n*, *180*, *181*, 182, 183, *183*, 197–8, 220, 242, 251, 252, 254, 255, 294, 298, 299, 308–9, 312, 325, 332, 335, 345
Feynman diagrams 176, 178–82, *180*, 252, 257, *257*, 258, 286, 345
field theory 2, 122, 177, 248–51, 260, 279, 320, 333, 345, 349
Fourier, Jean-Baptiste 31, 121, 145, 311–12

Galilei, Galileo 12, 16, 17, 18, 24
Galois, Évariste 34, 44–5, 47–8, 170, 337–8
gamma particle 243, 263, 345
Gamow, George 91, 93, 220, 243, 262–3, 337
gauge theory 250–54, 257–8, 260, 274, 282, 345
Gell-Mann, Murray 220, 255, 256, 257–8
Germain, Sophie 51

INDEX 355

Germer, Lester 97
Gibbs, Josiah 323
Glashow, Sheldon 181n, 259
gluons 257–8, 259
Goeppert Mayer, Maria 220, 244, 245
Goethe, Wolfgang von 208–10, 234
Gross, David 258
group 34, 38, 44–8, 50, 51, 250, 255, 256, 259, 268, 274, 279, 281, 300, 309, 329, 342, 345, 347

h 70, 71, 72, 73, 75, 80, 83–4, 89, 95, 108, 119, 200, 207, 285, 289, 297, 314
Hahn, Otto 236–8, 240
half-life 230, *231*, 232, 346
Hamilton, Sir William R. 12, 24, 25, 26–7, *27*, 28, 30, 32, 43, 45, 47, 108, 119, 123, 248, 249, 251, 279, 300, 308, 327, 337, 343, 346, 347
Hartree–Fock 166, 176, 177–8, 179, 187–9, 191, 204, 286, 346
Hawking, Stephen 331–3, 333n
Heisenberg, Werner 16, 32, 61, 64, 90, 91, 100–3, *101*, *103*, 104, 112, 115, 118–21, 120n, 122–3, 125, 128, 133, 135, 136, 143, 144, 145, 149, 150, 164, 209, 209n, 234, 241, 248, 250, 251, 296, 304, 331, 335–6, 343, 344, 347
Hermann, Grete 138, 163–4
hidden variables 142–3, 157–61, 163–4, 346
Higgs boson 259, 260, 271
Higgs, Peter 220, 253–4, 259–60
Hilbert, David 37, *38*, 337
Hilbert space 129, 133, 169, 174, 175–6, *176*, 180, 211, 302, 313, 319, 326, 327, 329, 346, 351
Holevo, Alexander 302
Hoyle, Fred 220, 264, 265
Hugo, Victor 100
Huygens, Christiaan 61, 269–70

imaginary numbers 27, 28, 108
isotope 229–32, 346

Jordan, Pascual 119, 122, 248
Josephson, Brian 285–6, 289, 296
Joule, James 172–3
Jozsa, Richard 161
junction 285–6, 289

Kadanoff, Leo 280
Kekulé, August 197–9, *198*, 282
Kelvin, Lord 230
Ketterle, Wolfgang 215
Kitaev, Alexei 294, 320
Kochen, Simon 163–4

Landau, Lev 34, 39–40, 43, 44, 91, 200, 280–2, 329, 337
laser 217–19, 301, 314
Lemaître, Georges 262
lepton 181n, 254, 259
local realism 138, 148, 149, 151, 157, 160, 346
locality 152, 157, 346
Loschmidt, Johann 172–3

Majorana, Ettore 320–1
Maldacena, Juan 330, 333n
Manhattan Project, the 240–3, 263, 305
matrix 28–9, *29*, 30, 32, 109, 347
matrix mechanics 62, 118–21, 122, 123, 125, 342, 343, 347
Maxwell, James Clerk 81, 173, 211, 323, 324–5
Meitner, Lise 220, 236–8, *236*, 240, 338
Mendeleev, Dmitri 38, 47, 54–5, 90, 126, 184, 185–7, *185*, 188, 228, 229, 244, 265, 338
Mermin, David 168, 168n
meson 246–7, 253, 255, 256
Millikan, Robert 80
Mills, Robert 220, 252–4, 258, 260
Montaigne, Michel de 163
MRI 7, 137, 285
Musil, Robert 111

neutrino 183, 229*n*, 245–6, 254, *257*
neutron 45, 96, 170, 186, 227, 228–30, *229*, 229*n*, 232, 234–5, 237, 243–6, 255, 257, *257*, 258, 265–7, 279, 305, 345, 346, 347
Newton, Isaac 2, 6, 12, 19, 20–26, *21*, 29, 35–6, 43, 45, 60–61, 73, 77, 83, 84, 95–6, 112, 120*n*, 122–3, 124, 125, 179, 184, 208–10, 209*n*, 225, 228, 260, 296, 323, 336
Noether, Emmy 34, 35–41, *36*, 46, 49, 127, 163, 203, 240, 250, 280, 336
nuclear fission 227, 236, 237, 266, 347
nuclear fusion 236, 263, 266, 347

Onnes, Kamerlingh 268, 282–3, 337
Oppenheimer, Robert 117–18, 220, 239–42, 255
orders of magnitude 74

path integral 123
Pauli, Wolfgang 28*n*, 34, 48, 49–50, *49*, 54, 91, 104, 115, 120, 122, 126, 129–30, 144, 145, 149, 166, 171, 177, 179–80, 187, 192, 241, 244, 246, 248, 252, 265, 323, 336–7, 343–5, 349
Pauling, Linus 166, 196
permutation group 45, 47–8, 170, 347
perturbation theory 248–9, 258, 345
photoelectric effect 75, 78–81, *79*, 84
Planck, Max 64, 65–7, *66*, 68, 69, 70–83, 85, 88, 90–91, 95, 108, 110, 119, 125, 144, 167, 200, 212, 216, 217, 285, 289–90, 331, 335
Planck's constant 69, 73, 88, 95, 108, 119, 125, 285, 289–90
Politzer, David 258
positron 127, 181, 250–1
Preskill, John 141–2, 294, 319, 332–3
proton 74, 94, 96, 137, 170, 184, 186, 189, 227–30, 229*n*, 232, 233, 234, 243, 244–6, 255–7, *257*, 270, 276, 279, 345, 346, 347, 348
pudding 64, 85, 86, 87, 88, 144, 226

quantum
  chromodynamics 254–5, 258–9
  computer 154, 216, 258, 286, 292–3, 294, 295–6, 302, 305–21, *313*, 326–7, 329, 332, 343, 344, 348, 349
  computing 162, 294, 301, 302, 312, 315, 316, 318, 326–7, 332, 334
  cryptography 294, 303–5, 308, 324–5
  electrodynamics 248, 250–2
  error correction 294, 305, 317–21, 328, 334
  Hall effect 281, 287, *287*, 288–91, 296, 320
  information 294, 301–4, 324, 327, 332
  monogamy 294, 303–4, 334
  simulation 298–301
  teleportation 161–2, *161*, 162*n*, 324
  tensor networks 182, 294, 315, 326–30, *328*
  tunnelling 116–18, *117*, 230, 285, 300, 350
quark 7, 45, 181–2, 183, 186, 255, 256–9, 261, 270, 271, 272, 276, 336, 345, 349
quaternion 26–30, 28*n*, 349
qubit 133–5, *134*, 137, 157, 160, 162, 162*n*, 169, 172, 174, 182, 295, 299, 302–10, 312, 314–16, 318–22, 326, 328, *328*, 329, 333, 342, 348, 349

renormalization 249, 252, 278–81, 328, 349
  group 268, 279, 281, 300, 309, 329
Röntgen, Wilhelm 221–2
Rutherford, Ernest 64, 86–8, *87*, 90, 220, 224–31, *225*, 233, 239, 249, 263, 336, 337

Salam, Abdus 254, 259
Schrieffer, Robert 283

INDEX   357

Schrödinger, Erwin 16, 32, 90, 104, 106–11, *106*, 112, 114–15, 118, 120–2, 125–6, 128, 133, 143, 144, 216–17, 234, 250, 251, 295, 336, 338
  Schrödinger equation 1, 107–9, *107*, 111, 114–15, 118, 120–1, 125–7, 133, 187, 188, 308–9, 344, 347, 351
  Schrödinger's cat 109, 138, 152–5, *153*, 156
Schwinger, Julian 220, 251
semiconductor 178, 205–6, 315, 349
Shor, Peter 294, 311, *311*, 312, 313, 319, 337
Specker, Ernst 163–4
spin 20, 28n, 75, 126–7, *126*, 129–32, 134–7, 141, 159, 162, 187–8, 189n, 193, *193*, 249, 252, 256, 260, 345, 350
standard model 38, 220, 259–60, 275, 276, 277, 292
Stevin, Simon 10, 12, 13–16, *13*, 18, 18n, 21, 24, 44, 59, 74, 116, 134n, 167, 260, 334, 336–7, 340
string theory 261, 271, 320, 321, 330, 333
strong nuclear force 245, 253, 254, 260, 247
superconductivity 44, 109, 253, 273, 281–6, 296
superposition 30–32, *31*, 52, 53, *53*, 54, 55, 60, 94, 99, 104, 108–9, 113, 115, 118, 132–7, *134*, 141, 148–9, 150, 152–5, 167, 170, 193, 198, 249, 256, 292–3, 302, 304, 314, 318, 319, 320, 334, 343, 346, 348, 349, 350, 351
symmetry 1, 2, 34, 35–55, 62, 73, 124, 170, 171, 181n, 187, 195, 203, 204, 244, 250, 252–3, 254, 255, 256, 261, 280, 281, 290, 301, 307, 311, 329, 337, 345, 350
  breaking 34, 39, 40–44, *42*, 50, 200–1, 201n, 216, 253, 259, 268, 273–4, 282, 284–5, 301, 350
Szilard, Léo 242, 324

Thomson, J. J. 85–6, *86*, *87*, 226, 336
Tomonaga, Shinichiro 220, 251
transistor 7, 134, 139, 166, 168, 205–6, 286, 299, 306, 349

uncertainty principle 2, 102, 103, 119, 122, 135, 143, 144, 145, 149, 150, 296, 304, 331, 343, 344, 350–1

van Leeuwenhoek, Antonie 59
Vidal, Guifré 328
von Klitzing, Klaus 268, 287–91
von Neumann, John 128–9, 156–7, 163, 224, 306–7, 317

wave function 23, 32, 47, 48, 104, 107, *107*, 108, 109, 111–17, 113n, 120, 124, 129, 132, 133, 136, 143, 148–9, 153, 166, 169–72, 174, *176*, 177–9, 182, 188–9, 281, 292, 342, 344, 345, 346, 347, 351
wavelength *67*, *68*, 74, 94–6, *95*, 99–100, 102, 116, 124, 125, 132, 207, 214, 284, 335, 343, 348, 351
weak nuclear force 227, 245, 254, 257, 260, 347
Weinberg, Steven 220, 254, 259, 275, 276–7, 337
Wheeler, John 325
White, Steven 328
Wieman, Carl 215
Wilczek, Frank 258, 291
Wilson, Kenneth 268, 279, 280–1, 337

Yang, Chen Ning 220, 252–4, 258, 260
Young, Thomas 81, 81n
Yukawa, Hideki 220, 246–7, 252, 253, 255, 256

Zeilinger, Anton 116, 160
Zoller, Peter 294, 313–14